South Asia and Climate Change

This book provides a comprehensive and interdisciplinary examination of the diverse aspects of climate change in South Asia. The region, home to almost 4% of the world's population, is under serious threat from climatic disasters. The volume underscores the urgency of addressing cataclysmic events related to climate change and their ramifications on the economy, agriculture and livelihoods of the region. The book discusses the reasons causing climate change as well as highlights normative and ethical considerations involved in the battle against climate change.

With case studies from India, Sri Lanka and Bangladesh, it explores issues such as extreme climatic events; energy use, fossil fuels, non-renewable resources and carbon dioxide emission in South Asia; internal migration and climate refugees; the ethical dilemma of sustainable development; technological advancements for extreme weather forecasts; and responses to climate change in South Asia. Highlighting the need for striking a balance between developmental imperatives and environmental sustainability, the chapters also show the North-South divide in the research agenda and policies on climate change and the global politics that underlie climate policies. The volume juxtaposes a scientific analysis of factors responsible for climate change with an analysis of the human cost of climate change from the perspective of social sciences. It discusses the challenges faced by developing countries while also offering recommendations and solutions.

This book will be of interest to scholars and researchers of climate studies, geography, public policy and governance, sustainable development, development studies, environmental studies, political studies, international relations, political economy, economics and sociology. It will also be useful to practitioners, think tanks, policymakers and civil society organisations working on environmental management.

Mausumi Kar is Associate Professor of Economics, Women's Christian College, Kolkata, India, and Guest Faculty of International Economics at University of Calcutta, India.

Jayita Mukhopadhyay is Associate Professor of Political Science, Women's Christian College, Kolkata, India, and former Guest Faculty, Department of International Relations, Jadavpur University, India.

Manisha Deb Sarkar is former Associate Professor and former Head, Department of Geography, Women's Christian College, Kolkata, India, and former Guest Faculty at University of Calcutta, Jadavpur University and West Bengal State University, Barasat, India.

'This volume on climate change in South Asia, edited by Kar et al., is a significant addition to the available literature on climate change in one of the most vulnerable and populous regions of the world. In particular, the emphasis on extreme weather conditions and recurrent natural disasters is an excellent and timely intervention. Strong evidence with analytical rigor makes it an invaluable reference to researchers and policy makers.'

Sugata Marjit, *Distinguished Professor, Indian Institute of Foreign Trade, Kolkata; Former Director, Centre for Studies in Social Sciences, Calcutta, and Former Vice-Chancellor, Calcutta University, India*

South Asia and Climate Change

Unravelling the Conundrum

Edited by Mausumi Kar, Jayita Mukhopadhyay and Manisha Deb Sarkar

Routledge
Taylor & Francis Group
LONDON AND NEW YORK

First published 2022
By Routledge
2 Park Square, Milton Park, Abingdon, Oxon OX14 4RN

and by Routledge
605 Third Avenue, New York, NY 10158

Routledge is an imprint of the Taylor & Francis Group, an informa business

British Library Cataloguing-in-Publication Data
A catalogue record for this book is available from the British Library

Library of Congress Cataloging-in-Publication Data
A catalog record for this book has been requested

ISBN: 978-0-367-47812-4 (hbk)
ISBN: 978-1-032-03670-0 (pbk)
ISBN: 978-1-003-04573-1 (ebk)

Typeset in Sabon
by Apex CoVantage, LLC

Contents

Figures

Tables

Contributors

Sanjib Bandyopadhyay is Deputy Director General of Meteorology and the Head of Eastern Region covering the states of West Bengal, Bihar, Jharkhand, Orissa, Andaman & Nicobar Islands and Sikkim at the Indian Meteorological Department, Government of India. He has 20 years of forecasting experience in different fields of meteorology, including tropical cyclones which strike the eastern coast, notably the recent cyclones like Fani, Bulbul and Amphan. He is the nodal officer of implementation of the flag ship project of Niti Ayog, 'Gramin Krishi Mausam Seva'.

Srijan Banerjee is Research Assistant at the Centre for Studies in Social Sciences, Calcutta, India. He completed his postgraduate studies in economics from the University of Calcutta, and his research interests are focused on environmental economics, macroeconomics and economics of growth. He completed a research project funded by ICSSR-ERC and significantly contributed to the 'District Human Development Report: Howrah, West Bengal', funded by Government of West Bengal.

Anindya Basu is Assistant Professor at the Department of Geography, Diamond Harbour Women's University, India. A rank-holder in the bachelor's and master's degree programmes in geography, University of Calcutta, she previously served at Women's Christian College, Kolkata, and was a UGC-JRF in the Department of Geography, University of Calcutta. She has interests in socio-political issues, gender studies and urban geography. She has published several research papers in reputed journals and has presented papers in national and international seminars.

Ganesh Kumar Das is Senior Forecaster of Eastern India posted at Regional Meteorological Centre, Kolkata, India. He completed his post graduation (gold medalist) in pure mathematics from University of Calcutta and Ph.D. in atmospheric science from University of Calcutta. He has published more than 20 research papers in journals and has 15 years of experience in weather forecasting. His area of interest is the forecasting of severe weather events, especially tropical cyclones and severe thunderstorms.

Manisha Deb Sarkar is former Associate Professor and Head, Department of Geography; Women's Christian College, Kolkata, India. She teaches postgraduate courses in geography at different state and national universities in West Bengal, India, and has nearly thirty-three years of teaching experience at the undergraduate level, including at University of Calcutta and Jadavpur University. She specialises in human and social geography, climatology, river and water resources and has a keen interest in historical geography. She has completed research projects funded by the UGC and Calcutta University as Junior Research Fellow and Principal Investigator, respectively. Her research articles have been published in journals, as chapters in books, in monographs and in international conference e-publications. She edited the book *Geo-Political Implications of Partition in West Bengal* and jointly edited the book *Human Rights Education*.

Sarbani Guha Ghosal is Associate Professor of Political Science at Bijoy Krishna Girls' College, Kolkata, India. She is an alumna of Presidency College, Kolkata. Her publications include *Civil Society: A Brief Kolkatan Perspective* and an edited book, *On Secularism*. She has written for *The Indian Journal of Political Science*, *Socialist Perspectives*, *Journal of West Bengal Political Science Association*, *Social ION* and so on and has contributed chapters in edited books.

Anwesha Haldar is Assistant Professor of Geography at East Calcutta Girls' College, Kolkata, India. She has also taught at the postgraduate level in University of Calcutta and Rabindra Bharati University, Kolkata. She graduated from Loreto College and was a UGC-SRF in the Department of Geography, University of Calcutta, from where she received her Ph.D. She has published many research articles in reputed journals, edited books, and presented papers in the fields of urban geomorphology, climatology, and environmental geography.

Mausumi Kar is Associate Professor of Economics, Women's Christian College, Kolkata, India. She also teaches international economics at the postgraduate level at the University of Calcutta, and her research interest is focused on international trade, economic integration and South Asia and various aspects of the textile and clothing industry of India. She has published in journals including *The World Economy*, *Contemporary Economic Policy*, *Journal of Policy Modeling* and *South Asian Journal of Macroeconomics and Public Finance*. She has presented at international conferences like UNU-WIDER and WEAI and authored the book *Indian Textile and Clothing Industry: An Economic Analysis* (2015), apart from contributing to several other edited volumes.

Nabendu Sekhar Kar is Assistant Professor in Department of Geography, S.M.H. Government College for Women, West Bengal, India. Previously, he served at Chandernagore Government College and was a CSIR-SRF in the Department of Geography, University of Calcutta. He specialises in geomorphology of humid tropics and RS-GIS. His works are focused on the geomorphological and climate-change issues of the Lower Bengal Delta. He has published research articles in several reputed journals and presented papers in conferences.

Sreya Maitra is Assistant Professor in the Department of Political Science, Maulana Azad College, Kolkata, India. She obtained her Ph.D. from Faculty of Arts, Jadavpur University. She has published papers on conflict and internal security in India and Sri Lanka, contemporary issues of domestic politics in India and Indian foreign policy. She has contributed articles to many journals and edited books.

Satyabrata Mandal is Assistant Teacher of Geography in a higher secondary school and received his Ph.D. in climate science from the University of Calcutta, India, in 2018. His research interests include climate variability and its implications in agrarian systems, climate change vulnerability in coastal islands, climate change projection, climate change and soil health and climate-resilient cropping systems.

Jayita Mukhopadhyay is Associate Professor in the Department of Political Science, Women's Christian College, Kolkata, India. She specialises in international relations and politics of South Asia, South East Asia and public administration. She is the author of the book *Role of ASEAN in Conflict Management: Potentialities and Pitfalls*, apart from publishing several research papers in reputed journals and presenting papers at seminars/conferences. Her popular articles have appeared in *The Statesman*, a national daily of India. She has acted as resource person in the World Congress of Political Science (Brisbane, Australia, 2018, and Poznan, Poland, 2016).

Ujjaini Mukhopadhyay is Associate Professor of Economics in Behala College, Kolkata, India. Her research interests include development economics, labour economics, gender economics and environment management. She has co-authored two books, published many articles in reputed journals and presented papers at conferences. She received the Emerald Literati Award from Emerald Publishing, UK, for Outstanding Paper in 2019. She is a regular reviewer of academic papers in journals and appears in newspapers and on television shows as an economics expert.

Gouranga Nandy is a media professional and development activist and currently works for the Dhaka, Bangladesh-based Bengali daily *Kaler Kantho* as Khulna Bureau Chief. He has served at several national dailies and magazines and published numerous news pieces, articles and columns focusing on Sundarbans mangrove forest and the coastal zone of Bangladesh in various national and international print media. He is also a contributor to BBC Bangla and the author of twelve books on the environment, the liberation war of Bangladesh, 1971, and land rights issues.

Ratnakar Pani is Principal of Prafulla Chandra College, Kolkata, India. He is a cost accountant and an associate member of the Institute of Cost Accountants of India. He is a life member of Indian Accounting Association and Research Foundation. His research interests are focused on energy management and environmental accounting. He has published articles in international journals and contributed chapters in edited volumes, apart from presenting papers in conferences.

Pradip Patra is Research Scholar of Climate Science at the University of Calcutta, India. He has qualified for CSIR JRF, UGC-NET and WBSET. He has presented his research articles in various seminars and also published his research articles in national and international books and journals. His research interests include climate variability, climatic hazards and its vulnerability assessment and impact analysis in rural areas and climate change projection.

Moushakhi Ray is Assistant Professor in the Department of Economics at Siliguri College, Darjeeling, West Bengal, India. Her research interests are in the areas of theoretical and empirical studies in international trade and the credit market, along with issues of sustainable development. Her major publications include 'Credit Constraints, Fragmentation and Inter-Firm Transactions' in *Asia-Pacific Journal of Accounting & Economics* (2014) and 'Export Profitability, Competition and Technology' in *International Review of Economics and Finance* (2017).

Lakshminarayan Satpati is Professor of Geography and Director, UGC-Human Resource Development Centre, University of Calcutta, India. His research and teaching interests include climatology, geomorphology, hydrology, environment, population and development and quantitative geography. He has several academic publications and administrative responsibilities at the Geographical Society of India, Kolkata, and Indian Meteorological Society (IMS), Kolkata Chapter, apart from being the Chairman, High Powered Committee to Restructure NATMO and editor of the *Indian Journal of Landscape Systems and Ecological Studies*. He has presented over 60 research papers and chaired sessions at national and international conferences.

Preface

The final shape of this volume was given by its editors at a time when the world is going through the excruciating experience of the COVID-19 pandemic, an unprecedented crisis faced by the global community. Experts opine that when we go about destroying our ecosystem through activities such as deforestation, mining and illegal hunting, we reconfigure the environment in a way that makes zoonotic transmission of viruses causing deadly diseases more rampant. Faced with the loss of habitats, animals are forced to venture into human settlements, thereby increasing the risk of transmission of potentially fatal pathogens. The loss of biodiversity and green cover of the earth, our only home, and concomitant climate change, happening at an alarming rate, is a wakeup call for every one of us. For a long time, the international community and particularly the UN Framework Convention on Climate Change (UNFCCC) has been working relentlessly to control the crisis globally. By February 2020, the Paris Climate agreement, which aims at keeping the increase in global average temperature to well below 2°C above pre-industrial levels, had been signed by all UNFCCC members. And yet there continues to be a huge gap between commitments and actions, and our planet continues to experience calamitous outcomes of climate change.

We, the editors and contributors of this volume, belong to a region where climate change is manifesting through cataclysmic incidents happening at an increasing frequency. Very recently, the eastern coast of India was ravaged by the super cyclone named Amphan, quite unparalleled in its intensity. Fortunately, as there has been considerable technological development in monitoring and forecasting extreme weather conditions in South Asia (an issue addressed in this volume, along with myriad other related equally important matters), millions could be moved to safer places before the cyclone hit the coast, thus minimising loss of lives. Perhaps no more arguments are needed to underscore the importance of studying the multidimensional impact of climate change in South Asia and its material and human cost and, more importantly, the attempts of South Asian states to respond to this challenge, adaptation strategies, the politics behind the policies adopted by different states, the identification of lacunae in all these efforts and finding ways of overcoming the shortcomings in the interest of ensuring the continuation of human civilisation.

The approach of this volume is consciously multidisciplinary, as it aims at breaking new ground in understanding and responding to the issue of climate change. The first section of this volume, offering a panoptic view of climate change scenario of South Asia, is presented as an appropriate background to subsequent studies. The second section delves deep into the hardship faced by the teeming millions of South Asians, whose lives are often torn asunder by devastating natural calamities. Contributors trace the origin of human migration precipitated by climate change back to the time of Indus Valley Civilisation and study the subsequent trajectories of migration both within the geographical contours of specific countries and across national boundaries. The socio-economic risks associated with such migration, and the resultant political tensions, are also explained. In the third section, the spotlight is primarily on India, the most populous as well as prosperous and powerful country of South Asia, so as to find out how the political class of India is pursuing the climate agenda at home and abroad and to what extent it is making use of the South Asian Association for Regional Cooperation (SAARC), encompassing all major South Asian countries, to promote cooperative action directed towards slowing down climate change. Two specific case studies, one on the experiences of Bangladesh in assessing the viability of shrimp-aquaculture to impede climate change in the Sundarban area, and the other on Sri Lanka's handling of the tsunami of 2004, provide valuable insight into the extent of awareness among those entrusted with the task of framing policies in South Asian countries and their preparedness and expertise in the area of disaster management and other related issues. Finally, adaptation strategies involving ethical considerations and technological advancements in controlling the outcome of extreme weather events over South Asia are described at length in the concluding section.

Our study is a cry for more vigorous attention and thorough study and research in the area of climate change, overwhelmingly viewed as one of the greatest threats facing humanity. It tries to answer many questions, while being fully aware that the questions are openended and the search for better answers should never stop.

We humbly thank the anonymous referees of Routledge whose valuable comments and suggestions on the manuscript helped us enormously in improving the quality of the volume. We record our appreciation for our contributors, who have readily acceded to our requests for several revisions and modifications of their chapters so as to facilitate our work of fine-tuning the book. Our family, friends and colleagues deserve our gratitude for their support and encouragement. We joyously mention Ms. Ujjayini Kar for her valuable technical assistance in arranging the entire manuscript. We must also duly acknowledge our camaraderie, which enabled the three of us to work as a team, helping and guiding each other at every step. We sincerely hope that the academia, students, researchers, policy-makers and activists engaged in studying the environment and dealing with the issue of climate change will find this volume useful.

Part I

Climate change in South Asia

Emerging issues and trends

1 Contemporary issues of climate change in South Asia

A synthesis

Mausumi Kar[1]

South Asia, occupying only 3.4 percent of the world's land area and carrying about 24 percent of the world's population,[2] has faced severe climatic events in recent times, and this has led to considerable loss of lives and property. While extreme climatic events and remarkable changes in weather patterns are not uncommon in many other parts of the world, the densely populated South Asian countries stand to witness significant adjustments in their socio-economic and political fronts owing to such rapid and irreversible alterations (Caesar and Janes, 2018). Importantly, however, the total area of South Asia and its geographical extent are not precise and unambiguous. Apart from peninsular South Asia, which was formerly a part of the British Empire, inclusion of other countries as integral to this region is a matter of argument. South Asia, in general terms, represents the southern region of the Asian continent, comprising the sub-Himalayan countries and the adjoining countries to the west and the east (ADB, 2010). Topographically, it is dominated by the Indian Plate situated south of the Himalayas and the Hindu Kush range. South Asia is bounded on the south by the Indian Ocean and on land (clockwise, from the west) by West Asia, Central Asia, East Asia, and Southeast Asia. South Asia includes the territories of eight neighbouring countries, to be precise, Afghanistan, Bangladesh, Bhutan, Maldives, Nepal, India, Pakistan, and Sri Lanka. The South Asian Association for Regional Cooperation (SAARC), embracing all such nations of South Asia, was established in 1985 to take coordinated initiatives for variant crises that the region faces.

It is perhaps well known that the countries of South Asia, the habitat of one-third of the world's poor, are experiencing huge economic, social, and environmental damages. This amounts to compromising the growth potential and poverty reduction efforts and consequent difficulties in attaining other Millennium Development Goals in an era of accentuated risks posed by global climate change. Countries in the region of Greater Himalayas – especially Bhutan, India (primarily the Northern part), and Nepal – are facing increased frequency and magnitude of extreme weather events, resulting in flooding, landslides, damage to property and infrastructure, devastation of agricultural crops, disruptions in hydropower generation, and negative impact on human health. The coastal areas of Bangladesh, India, the Maldives, and Sri Lanka are at high risk from projected sea level rise that may cause displacement of human settlements, salt water intrusion, and loss of agricultural land and wetlands and will generally suffer a negative shock on economic activities, including fishing and tourism. It is important to understand that some of these effects could be attributed to negative externalities generated from actions of neighbouring countries in the region, and therefore the scope for economic, political, and social adjustments and negotiations remains supreme for the current environment. There is no doubt that climate change poses the most formidable threat to the sustainable development of the South Asian countries. Observed vulnerability is related to issues like population growth, population density, poverty reduction, and so on and manifested in a rather complex manner. The challenge becomes even more pressing since the region exhibits a broad range of extremes: in topography, from the tallest mountains to the largest deltas and to islands and coral reefs; in precipitation, from arid lands to vast plains susceptible to erratic flooding; and in the environment, from glacial to tropical. To make matters worse, as Ying et al. (2017) have precisely noted, climate change projections for South Asia by the Intergovernmental Panel on Climate Change (IPCC) indicate that warming is likely to be above the global average (ranging from 1.5°C–4°C against the global average of 2°C). It also predicts that the monsoon rainfall is likely to increase (ranging from 4.4–13 percent) and that intense precipitation would be frequent. The low-income developing countries of South Asia thus face a huge challenge to deal with such multifaceted climate change impacts and to avoid greater social, economic, and environmental damages.

As a manifestation of these possibilities, climate change is already affecting the physical security of vulnerable communities in South Asia. Moreover, these impacts will be intensely felt in contexts where governance is already stretched (as identified by Vivekananda, 2011). In fact, the communities, which are insufficiently prepared to cope with the effects of climate change across the sub-continent, can suffer severely from political instability, food insecurity, economic weakness, and large-scale migration. It seems, however, that the responses to climate change are failing to effectively address the full range of secondary consequences, in particular in conflict-prone and conflict-affected contexts with scarce resources.[3] In the Terai region of southern Nepal, for example, when the Kosi River burst its banks in the summer of 2008, 240 people were killed, crops and infrastructure were destroyed, and 60,000 people were displaced in the worst flooding in five decades. These people resettled among communities that were themselves struggling to survive. Naturally, tensions between the host communities and flood victims quickly escalated and were further fuelled by political groups who used flood victims' unmet expectations for clean water and shelter to feed anti-government sentiments (Vivekananda, 2011). Hence, climate change and the consequent natural calamities can exacerbate the political instability that many countries in this region already suffer from both internally and at the cross-border levels.

This chapter addresses several of these issues, as discussed in the following chapters of this volume. It had best be admitted that any change in climatic conditions has diverse impact on an economy, some of which remains obscure owing to lack of political, legal, and economic interest in the matter. Historically speaking, in the view of Levitt and Dubner (2009), global warming was underplayed in the 1970s when the average temperature fell by a degree, until another decade, when great expansionary forces in the global economy changed the trajectory. The global south also undertook several economic reforms around this time, and the aspect of global warming became prominent. The bi-causality involved in the relationship between global warming and economic outcomes is, however, unmistakable. Burke et al. (2015) reflect that businesses associated with the previous level of emissions throughout the twenty-first century will lower the per capita GDP by 23 percent below what it would have been otherwise. Moreover, the impact of global warming in a particular region depends on the initial temperature: countries with an average yearly temperature greater than 13°C (55°F) will observe decreased economic growth as temperatures rise. For colder countries, warming will be an economic boon. This non-linear response creates a massive redistribution of future growth, away from hot regions and towards colder regions. Hence, the shock of climate change–induced temperature rise will be greater on poor countries because they are mostly situated in regions that are already warm.

The North-South divide in climate change research

In an article published in *Nature* on climate change, Blicharska et al. (2017) rightly argued that presently most of the scientific agreements and policy instruments developed under the UN Framework Convention on Climate Change (UNFCCC) are generated in the North, and it is the Northern countries that have set the global climate change policy agenda since the beginning. As a matter of fact, the Northern domination of science globally relevant to climate change policy and practice and lack of research led by Southern researchers in Southern countries may hinder development and implementation of bottom-up global agreements and nationally appropriate actions in Southern countries. Earlier, Jeffery (2014) has also argued in a similar fashion and apprehended that this kind of North-South divide may impact the least developed countries (LDCs) and small island developing states (SIDSs) like Sri Lanka the most, as they are among the most vulnerable to climate change but offer little evidence to support concerted action. It is worth mentioning here that we are following the country classification by World Bank Country and Lending Groups[4] as done by Blicharska et al. (2017), where the term 'North' or 'Northern' refers to those countries that are members of the Organization for Economic Co-operation and Development (OECD) or are classified as high-income economies by the World Bank based on the estimates of per capita gross national income. These are largely, but not exclusively, countries in Europe, North America, East Asia, and Australasia. The terms 'South' and 'Southern' as used here by default refer to countries classified as upper-middle income, lower-middle income, or low-income economies, which are mostly, but not wholly, located in the rest of Asia, Africa, and Latin America.

The North-South divide in research relevant to climate change policy and practice favours richer countries, which emit more carbon and are less vulnerable to climate change, and puts LDCs and SIDSs at severe disadvantages. Indeed, the southern countries have limited scope to question policy bias in the targets or the goals set by the richer countries and deviate from the appropriate measures to mitigate regional impact. If we look carefully at the global North-South divide in research on climate change (e.g. Pasgraad and Strange, 2013), many negative consequences are frequently observable. The imposition of carbon tax as a policy towards a better environment in future may be cited as an example of such bias. The mechanism of carbon tax and credit creates a distortion, the burden of which is borne disproportionately by the countries of the south owing to greater dependence on fossil fuel-intensive energy use. North-South research collaboration is also plagued by non-uniform interpretations of ethical standards of doing research in developing countries and by inequitable funding. After all, environmental research should be treated as a public good, and the burdens and benefits of doing research on environmental and climatic issues should be shared equally by the North and the South (Blicharska et al., 2017).

It can precisely be claimed that scientific advances are not the only yardsticks to measure the success of North-South research collaboration. Rather, the choice of identified priorities as areas of work, the sustainability of locally conceived actions, and the creation of capacity are becoming equally important as indicators of success. Long-term actions can be used to resolve ethical dilemmas and address inequities in research funding.

This volume, *South Asia and Climate Change: Unravelling the Conundrum*, is a modest attempt towards achieving that goal by identifying the extreme climatic events occurring in these regions; gauging their economic costs, including human displacement; examining the energy consumption patterns for regulating carbon emission levels; and finally recognising the region's commitment to minimising the vulnerabilities from political and ethical positions. The timing of the project could not be more appropriate, since every international leader of some worth has unanimously pointed out threats of climate change as the biggest issue facing our planet, and a sixteen-year-old teenager from Stockholm, Greta Thunberg, has become the biggest voice on this contentious issue, capturing the attention of world leaders and policy-makers through marshalling climate strikes throughout the world, receiving *Time* magazine's much-coveted 'person of the year' (2019) designation and many more awards in due recognition of her stupendous efforts.

The uniqueness of this project lies in the fact that this study also encompasses the role of regional, specialised meteorological centres in monitoring and forecasting the sudden occurrences of natural calamities like tropical cyclones and nor' westers, besides explicating scientific developments in the analysis of thermodynamic parameters and predicting and disseminating information in order to minimise loss of life and property.

Providentially, Asia has not been idle, and South Asia in particular is emerging as a global leader in responding to climate change aggressively. Through the South Asian Association for Regional Cooperation, South Asian countries have committed to a shared set of priorities in addressing climate change, including ambitious targets for clean energy and low-carbon technology development, as well as commitments to improve the resilience of communities and economies. The United Nations System Standing Committee on Nutrition (UNSCN) (2010) highlights how climate change further exacerbates the already unacceptably high levels of hunger and under-nutrition and proposes policy directions to address the nutrition impact of climate change for consideration by the 16th Conference of the Parties (COP) to the UNFCCC. The current negotiation process offers opportunities to identify and address some of the actions needed. Efforts to secure adequate water and food supplies are of critical importance.

The Coordinated Regional Climate Downscaling Experiment,[5] sponsored by the World Climate Research Programme, is exercised more vigorously and the flagship pilot studies driven by regional scientific communities are now carried out more frequently. It is worthy of mention that the global climate model (GCM) helps us in projecting future changes in the climate of the earth. The outcomes of such models primarily motivate the international community in taking decisions to lessen the adverse impact of climate changes. However, the adaptation strategies required to deal with these should at least develop more regional focus, if not local. This is where regional climate downscaling (RCD) has an important role to play by providing projections with much greater precision and more accurate representation of localised extreme events. Regional climate downscaling techniques, including both dynamic and statistical approaches, are being increasingly used to provide high-resolution climate information than is available directly from contemporary global climate models. This technical issue has been addressed at length in this volume. It elucidates the technological developments that are being carried out in order to equip the governments of South Asian countries, especially India, to restrict these undesirable events by proper monitoring/forecasting.

The global measures to combat climate change

In 1992, the United Nations Framework Convention on Climate Change was adopted as the basis for a global response to tackle the challenges posed by climate change. With 195 parties, the Convention enjoys near-universal membership. The ultimate objective of the Convention was to stabilise greenhouse gas (GHG) concentrations in the atmosphere at a level that would prevent dangerous human interference with the climate system. The Convention, also known as the Rio Earth Summit, entered into force on 21 March 1994, ninety days after the fiftieth country's ratification had been received. By December 2007, it had been ratified by 192 countries. As an extension to this treaty, and to consider further actions to address the climate change threat, the member states also negotiated a protocol to the Convention. The Protocol was named the Kyoto Protocol after Kyoto in Japan, where it was agreed upon by the member countries and countries of the former Soviet bloc to cut their emissions of greenhouse gases by an average of about 5 percent for the period 2008–2012 compared with 1990 levels. However, under the terms agreed in Kyoto, the protocol only enters into force following ratification by fifty-five parties to the UNFCCC. Hence, the Kyoto Protocol was adopted in Kyoto, Japan, on 11 December 1997 but not enacted or enforced until 16 February 2005. Emissions trading, as set out in Article 17 of the Kyoto Protocol, allows countries that have emission units to spare[6] to sell this excess capacity to countries that are over their targets. Thus, a new commodity was created in the form of emission reductions or removals. Since carbon dioxide is the principal greenhouse gas, the provision simply leads to trading of carbon. Carbon is now tracked and traded like any other commodity, formally known as the "carbon market". The clean development mechanism (CDM) is one of the flexible mechanisms defined in the Kyoto Protocol (IPCC, 2007) that provides for emissions reduction projects which generate certified emission reduction (CER) units that may be traded in emissions trading schemes (Marjit and Yu, 2018).

For quite a long time, the developing South, specifically the South Asian countries, has complained about the injustice done by international climate negotiations for placing a large burden on countries that have contributed relatively little to climate change, measured against Western industrial countries like the United States that were able to pollute for decades before being asked to curb their emissions. This made the challenge all the more complex for negotiators in Paris, where the Climate Conference had taken place, to reach a meaningful and actionable agreement for emissions reduction. The historic agreement to combat climate change and initiate actions and investment towards a low carbon, resilient, and sustainable future was agreed upon by 195 nations in Paris in December 2015 and is popularly known as the Paris Climate Deal. The Paris Agreement for the first time brings all nations into a common cause based on their historic, current, and future responsibilities, with a focus on curbing the global temperature rise this century to well below 2°C and to drive efforts to limit the temperature increase even further to 1.5°C above pre-industrial levels. The 1.5°C limit is a significantly safer defence line against the worst impacts of a changing climate. The Paris Agreement heralds a new bottom-up approach, with countries providing their own nationally determined contributions (NDCs) reflective of their ability and capacity to reduce the emission of greenhouse gases. However, the recent withdrawal of the United States from the Paris Agreement has once again ignited concern among countries about the economic impacts of taking measures to combat climate change. Nonetheless, some recent studies by Böhringer and Rutherford (2017) and Mani et al. (2018a) justifiably argue that the welfare loss of implementation of the Paris Agreement would be minimal for the United States as compared to that of the countries of European Union.

Vulnerability and responses of South Asia

A risk index, released at the annual climate summit in the Polish city of Katowice, in December, 2018, shows that countries in South Asia are among the most vulnerable globally to the impacts of climate change. India has been ranked the fourteenth most vulnerable nation in a list topped by Puerto Rico, while Sri Lanka, Nepal, and Bangladesh are among the top ten.[7] Climate change, or its most obvious form, global warming, is expected to cause living conditions to deteriorate for up to 800 million people in South Asia, a region that is already home to some of the world's poorest and hungriest people. A World Bank study by Mani et al. (2018b) looked at six countries of South Asia, where average annual temperatures are rising steadily and rainfall patterns are already changing. The study identified 'hot spots' where the deterioration is expected to be most severe and revealed that

> Hot spots are characterized by low household consumption, poor road connectivity, limited access to markets, and other development challenges.

In response, the developing countries of South Asia have reaffirmed their commitment to work together in addressing common challenges faced by this strategically important region. In 2007, the SAARC Council of Ministers adopted the SAARC Declaration on Climate Change, calling on SAARC leaders to collectively assess and respond to risks and impacts of climate change. In 2008, an action plan of three years was envisaged by the Environment Ministers of the SAARC countries at the Dhaka declaration on Climate Change that urges the international community to promote partnership and provide additional financing to support the actions. In 2010, the SAARC summit concluded with the Thimpu Declaration on Climate Change to set permissible levels for carbon emissions by the member countries. Other regional organisations are also taking supportive action. The Asian Disaster Preparedness Centre assists in enhancing capacities and skills to prevent damage, raising awareness to better prepare for disasters (ADB, 2010).

Ironically, despite the high risk for climate-related disasters owing to its unique topography and the poor air quality in many large cities, South Asia as a whole has a relatively light carbon footprint per capita. Though South Asia, especially India, ranks high among the large carbon-dioxide-emitting countries of the world, the region's carbon emissions rate stands at nearly 8 metric tons per capita (with India contributing only 1.5 metric tons) against the largest polluter, the United States, contributing nearly 16.5 metric tons during the same period.[8]

India's action plan, called a market readiness proposal, is the key to achieving climate goals. Under the Paris climate agreement, India plans to reduce its emissions intensity by 33–35 percent below 2005 levels by 2030. A nationwide carbon market could not only facilitate this but also promote social development. In addition to the immense economic costs associated with climate change (almost 5 percent of the global GDP per year), possible displacement of a significant quantum of population both on a temporary and permanent basis remains another pressing issue for the policymakers of the region. The growing effects of changing climatic conditions upon millions of people in South Asia who are being forced to leave their homes and migrate elsewhere is a serious matter of concern as of now. Human migration in this area is not restricted to within-country events but across borders, from the mountain to the plains, from the river banks to the interiors, and with considerable contemporary and future implications, economically and socially speaking. Apart from the standard risks associated with economic and political aspects, huge inter- or intra-state migration is often subject to the potential risk of adaptation failure and low resilience of the destination. Hence, as an important conduit of development, the worst affected need to get support for psychological adaptability based on the humanitarian, social, and political perspectives of the region.

This volume delves deep into the entire gamut of issues related to climate change in South Asia, its exacerbating dimensions, its correlation with economic growth and foreign direct investment (FDI), and, more importantly, its human cost involving the socio-psychological trauma faced by environmental refugees. Apart from highlighting the technological solutions for monitoring extreme weather conditions to minimise losses, the book also highlights normative and ethical policy considerations for the battle against climate change and for preventing future crises.

Structure of the book

The chapters in this edited volume address diverse issues that are imperative to analyse the conundrum emerging from and concerning climate change facets in South Asia. Snapshots of each of those themes are presented in this section for the reader's convenience.

Chapter 2 of this book summarises the threats of multifarious extreme climatic events (ECEs) that may occur in the South Asia Region, the home of one-fourth of the world's population, on account of rise in surface air temperature, precipitation anomalies, and other climatic incongruities. The chapter, with the aid of suitable climate models, predicts a temperature rise of 2°C to 4.9°C in this region at the end of the twenty-first century. Reviewing all possible impacts of ECEs, the chapter also suggests feasible options for location-specific adaptation and mitigation measures by resilience building, engaging stakeholders, involving local to global organisations, encouraging public and private partnerships, and above all local-level cooperation in coping mechanisms.

Chapter 3 presents a decomposition analysis of carbon dioxide emission of six South Asian countries using the logarithmic mean divisia index (LMDI) method for the period 1980–2015. In doing so, the authors have systematically recognised the role of the major contributing factors in changing emission levels apart from ascertaining the role of FDI in raising emissions. The study also investigates the extent to which changes in energy consumption patterns can regulate emission levels, based on which policy prescriptions are made.

Chapter 4 highlights the emerging issues of climate-change-induced human migration in the South Asian region dating back to Indus Valley civilisation, until recent times. The author articulates the factors behind such human migration, controversially called environmental refugees, often at a cross-border level. The chapter elucidates how socio-economic risks associated with adaptation failures and low resilience among those affected lead to rising political tension in the region.

Chapter 5 analyzes the nature and degree of correlation between economic growth, pollution, and migration in South Asia by employing a theoretical model, duly supported by a panel data analysis. It envisages that the promotion of export-led manufacturing growth with policy initiatives to control pollution in South Asian countries is a necessity to achieve sustainable economic development.

Chapter 6 focuses on the issue of region-wise vulnerability of India to persistent climatic events, including both slow- and rapid-onset drivers and their implications on escalating internal migration. The chapter delves into the phenomenon of the changing morphology of the river Ganga as a slow-onset event and its impact on human displacement in the Malda district of West Bengal, an eastern state in India. For the case of a rapid-onset climatic event, the 2015 floods in Tamil Nadu, a southern Indian state, and the consequent population displacement are considered. A portrayal of the precipitation anomaly is provided using secondary data. The trend of rapid and unplanned urbanisation is highlighted as a major cause of extreme climatic events.

Chapter 7 explores the intention and initiatives of the political class of India and their prudent action in dealing with climate change at home and, at the same time, gives a perusal of negotiations and joint ventures with its neighbours within SAARC to dispel the threats of devastations going beyond differences on many other fronts. In the light of the Paris Climate

Deal, this chapter offers valuable insight for averting the apocalyptic consequences of climate change in South Asia.

Chapter 8 evaluates the viability of shrimp aquaculture to impede climate change in the Sundarban area, the world's largest delta, known for its unique biodiversity, spread across both India and its neighbour Bangladesh, and the study makes use of the experiences of Bangladesh in this regard. This chapter also assesses in what way livelihood shift creates a greater rich-poor divide, further marginalising ecological refugees and how proper framing and implementation of fishery regulation can balance the situation. A study on the experiences of Bangladesh can provide better insight into the issue of sustainability of biodiversity, imperative for preventing the occurrence of biological desertification.

Chapter 9 illustrates a case study of environmental hazards, the tsunami that hit Sri Lanka, the small island country in South Asia, in December 2004. This is particularly relevant for South Asia, as it has a complicated security environment amidst momentous change in climatic conditions. This study exhibits the inter-relationship between political conflict and environmental disasters that is corroborated by analysis of the ground-level impact of the tsunami on the then-ongoing Sinhala-Tamil civil war in Sri Lanka. The comprehensive case study explores the international responses to the tsunami and the politics of aid and rehabilitation while trying to assess the lessons learnt in disaster management and levels of awareness regarding climate change at the policy level.

Chapter 10 examines the ethical dilemma of using the concept of sustainable development in mitigating unforeseen natural disasters and restraining environmental decay and challenges of climate change in South Asia. In light of environmental sustainability, the ideas of social justice, feminism and equality are discussed at length, with special reference to India.

Chapter 11 of this volume sheds light on the recent technological developments in monitoring and forecasting extreme weather events in South Asia, especially in India. By recognising the efficacy of better weather predictions in order to give communities more time to prepare for dangerous storms and to save lives and minimise damage to infrastructure, the study discusses the role of new communication networks, forecast system innovations and technologies (e.g. high-speed internet, wireless communication, digital climate database, Synergie workstations, nowcasting systems, ensemble forecasting systems). Also, the mode of dissemination of weather advisory and warning is been discussed in detail, with a special mention of Amphan, the recent super-cyclonic storm that had a disastrous impact on the land, mangroves and other physical resources of the eastern coast of India. The crucial technical aspects are substantiated by images of instruments and satellite images provided by the Indian Meteorological Department that are used to monitor the contemporary climate change phenomenon.

Finally, it is hoped that, given the coverage of this volume on a crucial geographical area, apart from students and general readers with an interest in the concerned area, the subject will attract academic and professional bodies working on climate change. Study groups and decision-making bodies in South Asian countries for environmental management and policies are also expected to find the book relevant and useful.

Notes

1 I wish to offer my sincere thanks to Prof. Saibal Kar, Centre for Studies in Social Sciences, Calcutta, Kolkata, India, and Prof. S. Marjit, Indian Institute of Foreign Trade, Kolkata, India, for their extensive guidance in completing this project.
2 Approximately 1.9 billion in 2018, based on the latest UN estimates.
3 See UN General Assembly Report (2009).
4 https://blogs.worldbank.org/opendata/new-country-classifications-income-level-2019–2020 accessed on7 June2020
5 www.cordex.org/
6 Amounting to the emissions permitted by the Protocol but not used by respective countries.
7 Global Climate Risk Index,2019, prepared by Berlin-based environmental organisation, German Watch.
8 Calculated in 2015 by US Department of Energy Carbon Dioxide Information Analysis Centre (CDIAC), mostly based on data collected from country agencies by the United Nations Statistics Division.

References

Asian Development Bank. *Climate Change in South Asia: Strong Responses for Building a Sustainable Future*. Asian Development Bank Publication. License:CCBY3.0 IGO, Manila, Philippines, November, 2010.

Blicharska, M., Smithers, R. J., Kuchler, M., Agrawal, G. K., Gutiérrez, J. M., Hassanali, A., Huq, S., Koller, S. H., Marjit, S., Mshinda, H. M., Masjuki, H. H., Solomopns, N. W., Staden, J. V., and Mikusinsky, G. 'Steps to Overcome the North-South Divide in Research Relevant to Climate Change Policy and Practice.' *Nature Climate Change,* 7(1) (2017): 21–27.

Böhringer, C., and Rutherford, T. F. 'Paris after Trump: An Inconvenient Insight.' CESifo Working Paper Series No. 6531. 2017. Available at SSRN: https://ssrn.com/abstract=3003930

Burke, M., Hsiang, S.M., and Miguel, E. 'Global Non-Linear Effect of Temperature on Economic Production.' *Nature* (November 12, 2015): 235–239.

Caesar, J., and Janes, T. *Ecosystem Services for Well-Being in DELTAS: Integrated Assessment for Policy Analysis*. Craig W. Hutton, W. Neil Adger, Susan E. Hanson, Md. Munsur Rahman and Mashfiqus Salehin Robert J. Nicholls (Eds.). Palgrave Macmillan, 2018. ISBN 978-3-319-71092-1ISBN 978-3-319-71093-8 (eBook). https://doi.org/10.1007/978-3-319-71093-8.

IPCC. 'Climate Change 2007: Impacts, Adaptation and Vulnerability.' In M. L. Parry, O. F. Canziani, J. P. Palutikof, P. J. van der Linden, and C. E. Hanson (Eds.), *Contribution of Working Group II to the Fourth Assessment Report of the Intergovernmental Panel on Climate Change*. Cambridge, UK: Cambridge University Press, 2007.

Jeffery, R. 'Authorship in Multi-Disciplinary, Multi-National North-South Research Projects: Issues of Equity, Capacity and Accountability.' *Compare: A Journal of Comparative and International Education*, 44 (2014): 208–229.

Levitt, S. D., and Dubner, S. J. *Super Freakonomics*. New York: William Morrow, 2009.

Mani, M., Bandyopadhyay, S., Chonabayashi, S., Markandya, A., and Mosier, T. *South Asia's Hotspots: The Impact of Temperature and Precipitation Changes on Living Standards*. Washington, DC: World Bank, 2018b.

Mani, M., Hussein, Z., Gopalakrishnan, B. N., and Wadhwa, D. 'Paris Climate Agreement and the Global Economy.' Policy Research Working Paper, No. WPS 8392. World Bank, South Asia Region, 2018a.

Marjit, S., and Yu, Eden. 'Globalization and Environment in India.' ADBI Working Paper Series, No. 873, September 2018.

Pasgaard, M., and Strange, N. 'A Quantitative Analysis of the Causes of the Global Climate Change Research Distribution.' *Global Environmental Change* 23 (2013): 1684–1693.

United Nations General Assembly. 'Climate Change and Its Possible Security Implications.' Report of the Secretary Genearal, United Nations, 2009.

United Nations System–Standing Committee on Nutrition (UNSCN). 'Climate Change and Nutrition Security.'*16th United Nations Conference of the Parties (COP16)*. Cancun, 2010.

Vivekananda, J. 'Practice Note: Conflict-Sensitive Responses to Climate Change in South Asia.' *IFP-EW Cluster: Climate Change Cluster*, October 2011. Available at: www.international-alert.org/ . . . /IFP Climate Change Conflict Sensitive Responses.

Ying, X., Bo-Taoa, Z., Jie, W., Zhen-Yua, H., Yong-Xianga, Z., and Jiaa, W. 'Asian Climate Change under 1.5–4 Degree Celsius Warming Targets.' *Advances in Climate Change Research*, 8 (2017): 99–107. Available at: www.sciencedirect.com

2 Extreme climatic events

A review of trends, vulnerabilities and adaptations in the South Asia Region

Satyabrata Mandal, Pradip Patra, Anwesha Haldar and Lakshminarayan Satpati

The South Asia Region (SAR) is considered one of the six sub-regions (namely Central Asia, East Asia, North Asia, South Asia, Southeast Asia and West Asia) of Asia characterized by unique geographical attributes and coastal configurations (Hijioka et al., 2014). The region comprises the eight contiguous South Asian Association for Regional Cooperation (SAARC) countries of Afghanistan, Bangladesh, Bhutan, India, Maldives, Nepal, Pakistan and Sri Lanka in the southern part of the continent (Figure 2.1) (World Bank, 2013; Sivakumar and Stefanski, 2014; Vinke et al., 2017; Raihan, 2018). It is bounded by the Arabian Sea (AS), the Bay of Bengal (BOB) and the Indian Ocean (IO) at the south. The region covers more than 5.2 M km^2 spread across eight countries, which is equivalent to 11.7% of continental Asia and 3.5% of the globe. This is the most densely populated geographical region, containing 1.9 billion people (39.5% of Asia), which is equivalent to about one-fourth (24%) of the global population, and it has the largest number (nearly 350 million) of the world's hungry people. The total population of the region is projected to be about 2.2 billion in 2050 (GWP, 2011; World Bank, 2013; Vinke et al., 2017). In accordance with the diverse physiographic settings, the region holds wide climatic variations form tropical (Bangladesh and Southern India) to subtropical (Northwest India) and tropical semiarid (central part) to alpine (the Himalayas). However, the typical climatic attributes of the region have been changing considerably in the recent decades as anthropogenic global warming and the concentration of greenhouse gases (GHGs) have increased rapidly. As the earth-atmosphere complex feedback mechanism is accelerating over spatio-temporal scales (local to regional), the various dimensions of the present climate system are changing rapidly, especially with the occurrences of extreme climatic events (ECEs) and their increasing frequencies and intensities across the South Asia Region. The unprecedented climatic change *vis-à-vis* increasing climate extremes (e.g. heat waves, floods, sea level rise, tropical cyclones, thunderstorms, prolonged dry spells, intense rainfalls, snow avalanches, severe dust storms, etc.), and their subsequent stress upon natural ecosystems are causing sustained degradation of natural resources (i.e. forests, soils, wetlands,

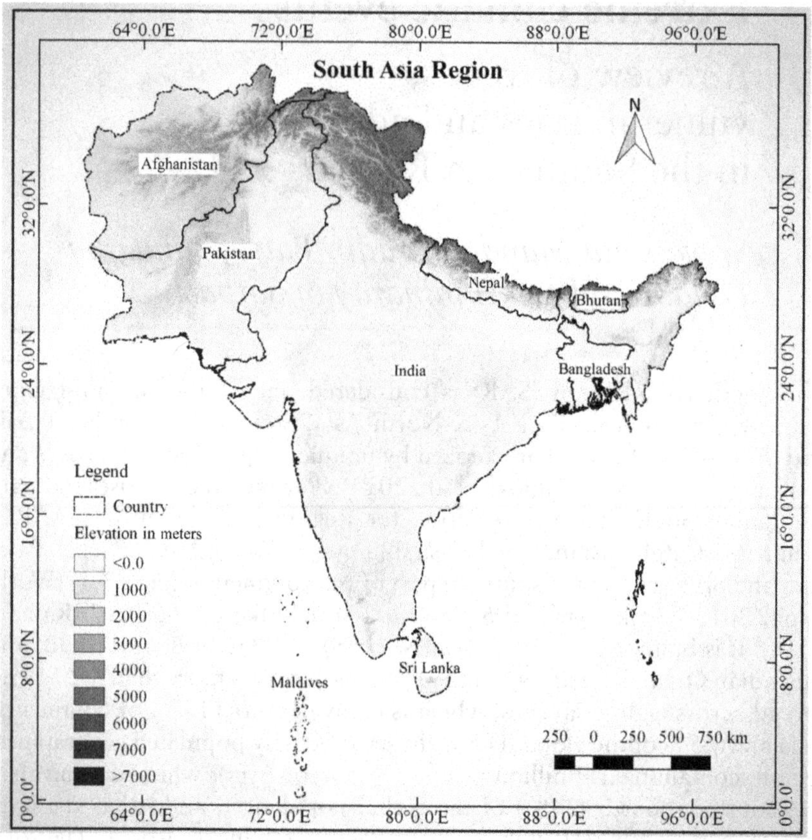

Figure 2.1 Location of the South Asia Region along with the spatial configurations of eight countries in the region

Source: Prepared by the authors using 7.5-arc-second (225-meter) GMTED 2010 product

Note: The international boundaries, coastlines, denominations and other information shown in any map in this work do not necessarily imply any judgment concerning the legal status of any territory or the endorsement or acceptance of such information. For current boundaries, readers may refer to the Survey of India maps.

rivers, aquifers, etc.) and human health as well as the security of livelihood in the region (Cruz et al., 2007). Robust evidence also exists in frequent occurrences of climate extremes at the regional and global scale in recent decades (Trenberth et al., 2015; Schoof and Robeson, 2016; Alexander and Arblaster, 2017).

The devastating floods of 2007 along the Ganga-Brahmaputra system affected more than 13 million people in Bangladesh alone, and another 20 million people were severely affected in the 2010 flood in Pakistan

(Dastagir, 2015; Dewan, 2015; Kreftand Eckstein, 2016; ADB, 2017). Likewise, India witnessed two unprecedented extreme flood events during August 2017 in the state of Assam affecting millions of people across 15 districts and another in the same month of 2018 over Kerala because of torrential (>53% above normal) downpour (Baynes, 2018; Mishra et al., 2018). Severe heat waves and droughts have affected vast areas of South Asia, including Western India and Southern and Central Pakistan in recent decades (Miyan, 2015). As part of perennially drought-prone regions, the countries of Afghanistan, India, Pakistan and Sri Lanka have repeatedly experienced onslaughts of drought during the past decades. Even Bangladesh and Nepal have suffered from frequent drought conditions in recent times (NDMC, 2006; Miyan, 2015). Moreover, the majority of climate projection models have warned of a decreasing trend of precipitation in summer and an increase in monsoon season (June to September) over the region (Christensen et al., 2007; Miyan, 2015). Hence, the region is highly vulnerable to ever-increasing occurrences of climate change–induced ECEs (ADB, 2017), and the less developed countries (LDCs) in the region (4 among the 49 LDCs of the world, viz. Afghanistan, Bangladesh, Bhutan and Nepal) would bear the maximum risks of climate change impacts because of their weak climate resiliency and low adaptive capacities (Kreft and Eckstein, 2016; ABD, 2017). Nevertheless, migration of coastal populations from high-risk and ECE-prone zones has been rapidly emerging as a new climate vulnerability in the region (ABD, 2017; Dastagir, 2015).

Of late, all 195 nations committed to limit global warming within 2.0°C at the United Nations Climate Change Conference (UNCCC) in Paris, so that the global mean temperature increase may be limited to 2.7°C above the preindustrial levels by the 21st century (Gütschow et al., 2015). If nations do not agree to the Paris pledges, the world's carbon budget covenant will be effective as early as by 2032 (Tollefson, 2015). In this regard, the South Asia Region is in the dual dichotomy of high risks of GHG emissions versus potential opportunities of economic development and observed versus projected impacts of anthropogenic global warming. However, it is really strenuous to keep balance between these issues in a region crowded by LDCs, and, accordingly, the challenge should be taken from the perspective of the sustainable development of the SAR. Therefore, efforts should be made on adaptive capacity building, and the strategies should involve preparedness programs (Mandal et al., 2017), urban and rural planning, national and social security schemes, proactive migration policy and so on. Crucial preconditions for success in the whole system should be long-term thinking and planning based on analysis of reliable data and modeling. Thus, the present chapter is aimed at providing a comprehensive overview of the available scientific findings on increasing climate extremes driven by escalating GHG emissions and observed and anticipated climate change over the South Asia Region.

Data and methods

The present chapter is mainly based on a comprehensive review of scientific literature available in the public domain and collection of relevant information through different local, national and international workshops and seminars on climate change, induced ECEs and their impacts in the LDCs of the SAR. A systematic flow chart of the review process and its academic outcomes are presented in Table 2.1. A review of the published literature – like peer-reviewed articles and essays in reputed journals, books and magazines

Table 2.1 Flowchart of the process of literature review followed in the chapter

Process stage	*Intended outcomes*
Selection of research topic ↓	• Formulation of the research questions • Defining the objectives • Defining the keywords pertaining to the research
Search of relevant literature ↓	• Searching the relevant sources of published materials • Use of Boolean operation • Preparing the inventory of relevant literature
Evaluate and select sources ↓	• Addressing the research questions/problems • Defining the key concepts and approach • Exploring the key theories, models and methods • Assessing the results and conclusions • Validation for confirmation, addition or challenging established knowledge, as applicable • Highlighting the key insights and arguments • Finding strengths and weaknesses • Taking notes and citing sources
Identify themes, debates and gaps ↓	• Screening for inclusion • Evaluation of the trends and patterns (in theory, method or results) • Identifying themes, debates, conflicts and contradictions • Selecting the pivotal publications • Finding research gaps
Categorizations of literature	• Statement of spatial and temporal scale • Observed change (temperature, precipitation, ECEs) • Projected change (temperature, precipitation, ECEs) • Country-wise scenarios • ECEs • Heat waves and drought • Floods • Cyclones • Chronological • Methodological • Theoretical

has helped in assimilating the relevant knowledge in this subject. Also, updated information from reputed newspapers, print and electronic media and blogs has been incorporated in this review. Moreover, information has been generated through consulting experts from research institutions, meteorological websites, rural and urban administrative bodies and non-governmental organizations (NGOs) for this chapter. The sources of information have been cross-checked to ascertain they meet acceptable standards in the scientific community.

It should be noted that most of the scientific publications on ECEs for the SAR have smaller geographical dimensions and are mainly country specific. But climatic phenomena are, in many occasions, cross-border in nature, such as cyclones occurring in India and Bangladesh, droughts in Pakistan and India and so on. So literature published at regional scale to cover the various types of ECEs together for the entire region is quite rare. Also, scientific studies based on reliable databases generated through updated technologies and their analysis using robust models are only contemporary developments, so publications are very few.

Observed and projected climate change in South Asia

Observed temperature change

In their 5th Assessment Report (AR5), the IPCC (2014) documented unprecedented warming across the globe. Krishnan and Sanjay (2017) reported a temperature increase of about 0.85°C globally during the last century (1880–2012). Similarly, the mean annual temperature in most of the SAR has also increased in the past century, which may have substantial impact on the climate system (UNFCCC, 2007; IPCC, 2014) of the region (Vinke et al., 2017). The annual average temperatures over the SAR, particularly in the Indian landmasses, have also increased by 0.6–1°C during the last century (1901–2010), where the maximum changes occurred in the post-monsoon (0.79–1°C) compared to monsoon (0.43–1°C) seasons (Krishnan and Sanjay, 2017). The IPCC (AR4) also reported that the number of cold days and nights has decreased, while that of warm days and nights as well as heat wave frequency has increased since the middle half of the 20th century across most of the Asian regions, including South Asia. Thus, climate change will have a definitive impact on the process of sustainable development in most of the LDCs in the SAR, as it create pressures on natural resources, as well as the natural environment, associated with rapid urbanization, industrialization and economic development (Hijioka et al., 2014; Vinke et al., 2017). Similar observations have been reported by researchers from different countries; however, their dimensions are not univocal throughout the region. In Pakistan, a significant warming trend for mean temperature was observed by 0.57°C in the past century (1901–2000), which was rather lower than the SAR average of 0.75°C (ADB, 2017). The

Global Climate Change Impact Study Centre (GCISC) revealed that, during the period of 1951–2000, majority of the agro-climatic zones – like the Baluchistan plateau (Table 2.2) and Central and South Punjab – had experienced warming trends (Akram and Hamid, 2015). Ahmed and Suphachalasai (2014) also reported an accelerated hike of 0.47°C during 1961–2007 in Pakistan, where the highest increase was recorded in the winter season, ranging 0.52–1.12°C. The hyper-arid plains, arid coastal areas and mountain regions of Pakistan are experiencing warmer (by 0.6–1.0°C) weather, where the warmth has increased significantly by 0.5–0.7% over the southern half of Pakistan (Table 2.2). This is definitely in agreement with the steady warming over the SAR in the past half century, though most of that is felt in winter and in the post-monsoon months (Anjum et al., 2005). Owing to its complex mountainous topography, Afghanistan has a unique type of climate which has significantly changed since the past few decades, particularly since 1960. Mean annual temperature has increased by 0.13°C/decade^{-1}, and the frequencies of hot days and nights have registered an increasing trend throughout the year (Savage et al., 2009; Ghulami, 2017). In their study, Aich et al. (2017) reported that the warming over Afghanistan as a whole increased by 1.8°C during 1951–2010, while the maximum was recorded as 2.4°C in the east, and 0.6°C in the Hindu Kush region (Table 2.2). Further, significant warming was also recorded in the glaciated area of Badakhshan in the northeast, with a measure of 0.3–0.7°C. In spite of its negligible contribution (about 0.025%) in GHG emissions compared to developed countries, Nepal is also warming at a faster rate (0.06°C year^{-1}), even in the glaciated Himalayan region as compared to the global average (Gurung and Bhandari, 2009; MOE, 2010; MoSTE, 2015a; Anup, 2018). A number of studies have documented significant warming across Sri Lanka in the past few decades. Anomalies of surface air temperature over Sri Lanka (1869–1993) suggest a prominent increase by 0.30°C century^{-1} (Rupakumar and Patil, 1996), and the mean surface temperature has increased by 0.016°C year^{-1} (Chandrapala, 1996) during 1961–1990. Moreover, using long period data, De Costa (2008) found significant warming in all climatic zones that exceed the global average rate of warming, and the warming trend has accelerated in recent years (1987–1996) by 0.025°C year^{-1} (Table 2.2) (Zubair et al., 2010; Esham and Garforth, 2013).

Projected change of temperature

Based on Coupled Model Inter-comparison Project Phase 5 (CMIP5) simulations under all four representative concentration pathway (RCP) scenarios, the IPCC (AR4) projected that the mean annual surface temperature would exceed 2°C as compared to the late 20th-century baseline – across the majority of land areas in the SAR – during the mid-20th century under RCP8.5, which would range from >3°C over South Asia to >6°C over the high latitudes in the late 21st century (Hijioka et al., 2014). However, under the RCP2.6 scenarios, the ensemble mean showed a less than 2°C increase above the late 20th-century baseline in both the mid- and late 21st century, though exceptional change (2°C and 3°C) occurred in the high latitudes. Furthermore, all the emission scenarios revealed increasing warming of sea surface temperature (SST) in tropical and subtropical oceans (IPCC, 2014). Analyzing the multi-RCM ensemble of CORDEX South Asia data, Krishnan and Sanjay (2017) projected that the mean surface temperature, relative to the reference period of 1976–2005, would increase by <1°C (under RCP2.6) over most of the Indian landmasses (e.g. Tamil Nadu in the south and Jammu and Kashmir in the north). The near-future (2021–2050) scenarios have exhibited a similar range of uniform changes (of <2°C) across the Indian subcontinent under the scenarios of RCP4.5 and RCP8.5 simultaneously. Warming across all of India has already been projected to exceed 3°C under the RCP8.5 scenario by the end of the 21st century, while higher change has been projected (that would exceed 4°C) in the semiarid north-west and north India (Krishnan and Sanjay, 2017). The mean annual temperature in Pakistan has also been projected to rise between 4.3 and 4.9°C by 2085, and such warming would be higher in the northern parts as compared to the southern parts the country (Akram and Hamid, 2015). Nevertheless, the projected mean temperature increase in Pakistan is expected to be higher than the global average (Chaudhry et al., 2009; Ahmed and Suphachalasai, 2014; ADB, 2017). NEPA and UNEP (2015) also projected similar above–global average warming for Afghanistan with respect to the baseline of 1986–2006. The upbeat scenario of RCP4.5 is that the temperature will increase by 1.5°C until 2050, followed by a period of stabilization and then additional warming of 2.5°C by the end of the 21st century (Table 2.2). The RCP 8.5 scenarios, conversely, reveal a projected increase of about 3°C and 7°C by the middle and the end of the 21st century, respectively. The spatial pattern of warming showed a wide diversity in the near future (2021–2050), with and anticipated range between 1.5 and 1.7°C in the central highlands and the Hindu Kush region and also 1.1–1.4°C in the lowland areas (Savage et al., 2009; NEPA and UNEP, 2015). The estimated temperature for Nepal also exhibited quick warming of 1.4°C by 2030, 2.8°C by 2060 and 4.7°C by 2090 (Table 2.2) (NCVST, 2009; GWP Nepal, 2015). In Bhutan, the warming is said to be higher across inter-mountainous valleys compared to the northern and the southern parts, and the projected peak temperature for the country may reach 3.5°C by the 2050s (Table 2.2) (Tse-ring et al., 2010; State of Climate Change Report for the RNR Sector, 2016). Therefore, warming of >2.5°C in the 21st century over South Asia may considerably affect the lifesaving agricultural system with the associated food security for the resource-poor farming community and the marginal people of the glaciated areas, with fluctuating annual melt water and adverse effects on natural vegetation, human health and the overall natural ecosystems of the region.

Observed trends and anomalies in precipitation

In the past century (1900–2005), precipitation in Asia increased significantly across the north and central regions but declined in South Asia (IPCC, 2007; Hijioka et al., 2014). Most of the climatic studies in the region are limited to trend analysis because of insufficiently long period-observed precipitation records in major portion of the SAR in the 20th century. Seasonal rainfalls exhibited inter-decadal variability with a declining trend and more frequent deficit monsoon rainfalls of uneven distribution over the region (Cruz et al., 2007; IPCC, 2013b). Not only that, the recurrence of heavy precipitation events has increased (Mandal and Choudhury, 2014); conversely, light rain events have decreased over the region. The frequencies of mid-monsoon breaks as well as monsoon depressions, however, remained constrained within decreasing seasonal rainfall occurrences, though the increase in extreme precipitation events (EPEs) happened in place of weaker rainfall events over the central Indian region, including other areas (IPCC, 2013). The observed increase in frequency of the majority of EPEs (Gautam, 2012; Gautam et al., 2009; Vinke et al., 2017) occurred over the west coast and central and northeast India (Ajayamohan and Rao, 2008; Goswami et al., 2006; Singh and Sontakke, 2002; Vinke et al., 2017). Furthermore, short-spell droughts also increased across India (Deka et al., 2012), possibly because of global warming, which favored a greater degree of likelihood of heavy floods as well as short-spell droughts. Thus, these overwhelming changes will gradually pose serious threats to agriculture, fisheries and allied sectors in the near future (Vinke et al., 2017). Annual precipitation in Sri Lanka also decreased by 144 mm (7%) during 1961–1990 compared to 1931–1960. Researchers (Jayatillake et al., 2005; Esham and Garforth, 2013) had already warned of a declining trend of annual rainfall for the period of 1931–1960, estimated as 2005.0 mm, which further declined to 1861.0 mm during1961–1990 in Sri Lanka (Table 2.2). Also, volume rainfall and number of rainy days have both declined in higher-elevation regions (Herath and Ratnayake, 2004; De Silva and Sonnadara, 2009). The Asian Development Bank (2017) documented a 10–15% decrease of rainfall in winter and summer seasons in Pakistan in its arid plains and coastal areas; however, it has increased exceptionally by 18–32% in the core monsoon region of the country (Table 2.2). Also, an extreme water crisis has been observed during seven strong El Niño events in the last 100 years that brought a drastic shortage of rainfall of 17–64% of the country. It was also observed that only a 6% rainfall shortage can increase irrigational water requirements in Pakistan by up to 29% (Akram and Hamid, 2015). Neither the annual nor the monthly precipitation in Bhutan follows any systematic pattern or trend across the country (Table 2.2) (Tse-ring, 2003). Savage and others (2009) analyzed that annual rainfall over Afghanistan had decreased by 2% decade^{-1} in the past fifty years, and recent (1998–2001) droughts (Table 2.2) might directly be correlated with La Niña conditions in the Pacific.

Table 2.2 Summary of observed and projected changes of temperature and precipitation across the South Asian countries

Country	*Changes in temperature (observed and projected)*	*Changes in precipitation (observed and projected)*
Afghanistan	Observed mean temperature increased by 1.8°C (above global rise), which is projected to rise by 1.7–2.3°C and 2.7–6.4°C under RCP 4.5 and 8.5 by 2050 and (2100), respectively (Aich et al., 2017).	Annual precipitation in most parts of the country decreases from 5–20%. Precipitation in MAM projected to decrease by 5%, and in OND, it will increase by 5–10% by 2050 (Aich et al., 2017; Doosti and Sherzad, 2015).
Bangladesh	Mean temperature increased by 0.20°C decade^{-1}, significant increase observed in northern hill area, south-east and southern coastal region. Projected temperature change lies between 0.5 and 2.1°C in 2050 compared to 1961–1990 (Rahman et al., 2012; Rahman and Lateh, 2015).	Monsoon precipitation increased by 7.5 mm decade^{-1}, while, pre- and post-monsoon rainfall decreased by 7.5 mm decade^{-1} and 5.5 mm decade^{-1} accordingly. Annual and post-monsoon precipitation is projected to change by +5.5% and –13.3% in 2020 (Rahman and Lateh, 2015).
Bhutan	Summer and winter temperature raised by 1–2°C from 2000. Mean air temperature projected to increase by 1°C and 2°C in 2039 and 2059 (Bhutan Climate Change Handbook, 2016)	Observed precipitation shows no significant deceasing trend. Although the trend is not significant, overall annual precipitation is projected to increase, while for winter, it is projected to decrease by 20–30% in 2050 (State of Climate Change Report for the RNR Sector, 2016).
India	Long-term (1951–2010) maximum, mean minimum and mean temperatures showed a significant increasing trend. Mean temperature has changed in most parts of the country by 0.1°C decade $^{-1}$. Most of the models showed change of temperature by as much as 2.4–4°C by the end of the present century (Rathore et al., 2013; Bal et al., 2016).	Long-term (1951–2010) trend of annual, monsoon and post-monsoon precipitation exhibited non-significant increasing trend (0.4–21.3 mm decade^{-1}) to decreasing trend (0.5–32.6 mm decade^{-1}) in most parts of the country. Projection showed an increasing trend of 15–24%; precipitation extremes will increase in peninsular India and north-west India (Rathore et al., 2013; Bal et al., 2016; Woo et al., 2018)

(*Continued*)

Table 2.2 (Continued)

Country	*Changes in temperature (observed and projected)*	*Changes in precipitation (observed and projected)*
Maldives	No significant observed trend has been found. Projected daily maximum temperature is projected to increase by 1.5°C by 2100 (NAPA, 2006).	Annual precipitation showed increase from 1990. Limited projection studies. Maximum daily precipitation is 180 mm in 100-year return period, but to increase twofold in 2050 (NAPA, 2006).
Nepal	Annual and seasonal maximum temperature increase by 0.56°C decade^{-1}, while minimum temperature by 0.02° decade^{-1}. Temperature projections estimated a rise of 1.4°C by 2030, 2.8°C by 2060 and 4.7°C by 2090 (GWP Nepal, 2015; DHM, 2017).	Significant increasing trend of rainy day frequency observed, while wet day frequency decreased. Recent future (2020–30) precipitation projected to decrease, while for the distant future, it has projected to increase (DHM, 2017; Bajracharya et al., 2018; Rajbhandari et al., 2018).
Pakistan	Increasing temperature noticed in Baluchistan, South Punjab, during 1951–2000, while other regions maintained a decreasing trend. Annual temperature will increase by 3.4–4.7°C in 2085; the northern part followed a more warming trend compared to the southern part (Akram and Hamid, 2015).	Observed annual precipitation reduced by 6%, while monsoon precipitation increased by 25% in the 20th century. Variability and shifting in precipitation change is projected in the north-east by 2050 and north-west by 2100. (Akram and Hamid, 2015).
Sri Lanka	Warming temperature trend observed as 0.16°C decade^{-1} in the last century compared to global 0.13°C decade^{-1}. Warming of 2.5–4°C has been predicted by 2100 compared to 1960–1990. (Chandrapala, 1996; Kumar et al., 2006).	Studies found a decreasing trend of 7% in 1960–1990 compared to 1931–1960. Both the recent (2030) and distant future (2050) precipitation characteristics of the 10th percentile showed decreasing trends, while median and 90th percentile showed increasing trends with 8–10% and 44–48% in the RCP 4.5 and 8.5 scenarios, respectively (Eriyagama et al., 2010).

Source*:* Authors' presentation

Abbreviations used: MAM: March-April-May; OND: October-November-December; RCP: Representative Concentration Pathway

Precipitation projection over the South Asia Region

By analyzing five bias-corrected global climate models (GCMs), Vinke et al. (2017) projected that the mean annual precipitation will increase under both the RCP 2.6 and 8.5 scenarios over most of the SAR (Cruz et al., 2007; Hijioka et al., 2014). Even though the precipitation projections may have large uncertainties, this considers the temperature-dependent energy balance model (Esham and Garforth, 2013; Vinke et al., 2017). Western Pakistan remarkably showed a reverse pattern of precipitation behavior. The annual mean rainfall will experience about a 30% increase, particularly in areas stretched between the northwest and southeast coasts of the Indian peninsula under the RCP 8.5 scenario. Moreover, the percentage change in summer monsoon rainfall considerably contributes to the annual precipitation. On the contrary, in Pakistan and the central and northern regions of India, the winter December-January-February (DJF) precipitation reveals a relative decline. Inter-model uncertainty is observed for rest of the regions under the RCP 8.5 scenario (Table 2.2). Thus, the findings are quite consistent with the IPCC AR4 (CMIP3) models that the south-west monsoon (SWM) season will be wetter, while the post-monsoon season will be drier. The RCP 2.6 scenarios showed large inter-model uncertainty for winter rainfall over most of the regions of India. Researchers from Sri Lanka (Cline, 2007; Cruz et al., 2007; De Silva et al., 2007) also reported similar predictions of increasing mean annual rainfall across the country, while others (Ashfaq et al., 2009; Esham and Garforth, 2013) projected a decrease in those (Table 2.2). Similarly, a downscale of two regional climate models (HadCM3 and CSIRO) suggested comparatively higher increases in SWM than north-east monsoon (NEM) rainfall across Sri Lanka (Kumar et al., 2006). Furthermore, it has been assumed that, rainfall in the *maha* cultivation season (September–March), considered with the NEM, will decrease by 9% to 17% (De Silva et al., 2007). Rainfall projections over Pakistan do not indicate any systematic patterns except an increasing trend over the Upper Indus Basin and a decreasing trend in the Lower Indus Basin (Zahid and Rasul, 2012; Iqbal and Zahid, 2014; Ahmed and Suphachalasai, 2014). The State of Climate Change Report for the renewable natural resources (RNR) sector (2016) documented that the climate models of ECHAM5 and HadCM3Q projected a progressive and steady increase in annual mean precipitation over Bhutan for the period of 1980–2069 (Table 2.2). The winter rainfall is expected to decrease up to 20–30% in the north-east and south-west parts of the country by 2050; however, the country as a whole will experience a significant overall increase in precipitation (Tse-ring et al., 2010).

Extreme climatic events in the region

Heat wave and drought events in South Asia

Most of the climate models projected an incline of global mean temperature by 5.5°C, as well as decline of post-monsoon precipitation by the end of 21st century over the SAR (Christensen et al., 2007; Cruz et al., 2007; Handmer et al., 2012; Stocker et al., 2013; Hijioka et al., 2014; Miyan, 2015; Mazdiyasni, 2017). This, in turn, increases the intensity of heat waves and droughts across the world (Hansen et al., 2012; Shi et al., 2015; Mazdiyasni, 2017), which should have a greater impact on the SAR, more specifically in low-adaptive, highly populated LDCs like Afghanistan, Bangladesh, Bhutan and Nepal (Diffenbaugh and Scherer, 2011; Mazdiyasni, 2017). Researchers (Sivakumar and Stefanski, 2014; World Bank, 2013; Miyan, 2015; Bandyopadhyay et al., 2016; Mazdiyasni, 2017; Vinke et al., 2017; Krishnan and Sanjay, 2017; Mani et al., 2018), however, have repeatedly reported such increasing extremes in the SAR. The quantifications of such increasing heat-related extreme events and their exposures and vulnerabilities (climate change risk), even with the uncertainties in their predictions, are now being documented widely across the studied region. Also, their widespread adverse impacts on natural ecosystems, location-specific economies and human wellbeing are well documented. The extreme positive departure (> +5–7°C) of maximum surface temperature from the long period average (LPA), characterized by the sudden occurrence and persistence of extremely high anomalous temperatures for several (>5) days is considered a heat wave/extreme (Steffen et al., 2014; WMO, 2015; Bandyopadhyay et al., 2016). Comparative studies of the occurrences of heat waves/extremes for 2°C (RCP 2.6) and 4°C (RCP 8.5) scenarios are substantially limited across the SAR. However, Vinke et al. (2017) have analyzed and compared events under these scenarios and documented them scientifically. The 4°C (RCP 8.5) global projection warned for a prominent increase in the frequency of warmer boreal summer months (5-sigma) compared to the LPA over the Indian subcontinent (in the Indian peninsula, Bhutan and parts of Nepal). They have even determined (by comparing multi-model means: 5-sigma) that there is a 60% chance of summer months getting hotter by the 21st century, though there is larger uncertainty about the timing and magnitude of the increase in the frequency of heat events (Table 2.3). Apart from this, there is a >50% and >90% chance (RCP 8.5; 3-sigma) of hotter summer months (MAM) in the northern and southern parts of the region by the end of this century. The RCP 2.6, on the contrary, predicted a rise in temperature only for the southernmost tip of India and Sri Lanka. All the CMIP5 models projected warm spells (>90th percentile) that will lengthen to 150–200 days under 4°C global scenarios, while it will be 20–45 days under 2°C (RCP 2.6) at the regional scale (Sillmann et al., 2013; Vinke et al., 2017). Decreasing precipitation in the post-monsoon season often plays a crucial role in the formation and intensification of heat extremes and droughts (Dai et al., 2004; Bandyopadhyay et al., 2016). Mani et al. (2018) reported that megacities like Dhaka, Karachi, Kolkata and Mumbai (home to >50 million people) are considered the hot spots of the incidences of climatic extremes. Even in 2015, the region lost more than 3,500 people (Mani et al., 2018), while in India alone, heat waves have caused 22,562 deaths since 1992 in different states (Table 2.2) (NDMA, 2016; Bandyopadhyay et al., 2016; Krishnan and Sanjay, 2017).

Drought risks and vulnerabilities

The increasing frequency and intensity of droughts in the tropical and subtropical areas of the region has become more common since the 1970s (Sivakumar and Stefanski, 2014; IPCC, 2007). Decreased precipitation coupled with increased temperatures in the post-monsoon season enhances evapotranspiration (Mandal and Choudhury, 2014), contributing to more regions experiencing droughts, especially in the LDCs that suffer from the maximum risk of loss of life and natural resources. Moreover, it is evident that El Niño episodes weaken the monsoon rains over Southeast Asia, particularly across the Indian sub-continent, which turn into severe and prolonged droughts across the country. It has been estimated that around 330 million people lost their resources due to drought, amounting to nearly US$3 billion in 2015–2016. Droughts with varying magnitudes have been discussed in detail and reported by researchers from different parts of the region, such as Dastagir (2015), Miyan (2015) from Bangladesh; Sivakumar and Stefanski (2014), Bandyopadhyay et al. (2016), Krishnan and Sanjay (2017) from India; Zahid and Rasul (2012), Ahmed and Suphachalasai (2014), Iqbal and Zahid (2014), Akram and Hamid (2015) from Pakistan; Herath and Ratnayake (2004), Imbulana et al. (2006) from Sri Lanka; and Savage (2009), WFP and UNEP (2016), Ghulami (2018) from Afghanistan (Table 2.3).

Extreme precipitation events and flood risks

Almost all the CMIP5 models and scenarios have projected an increasing trend in extreme precipitation events in the summer monsoon (IPCC, 2013a) season across Southern Asia. The inter-seasonal and annual precipitation also deviated sharply (IPCC, 2013a). Nevertheless, it is difficult to project an EPE, as it occurs from complex and dynamic processes at short timescales (ADB, 2017). Climate models (CMs) have limited success in identifying the monsoon mechanisms accurately; however, updated CM outputs show improved efficiency in capturing observed monsoon circulation with significant uncertainties (IPCC, 2013a). The LPA observations exhibited increasing frequency of EPEs, while light rain spells have decreased over the region (IPCC, 2013a). The rising temperature that has been already projected for the region leads to higher evaporation rates, due to the fact that the water-holding capacity of the atmosphere increases by 7% degree^{-1} of warming (Zhang et al., 2007; Fischer and Knutti, 2015), leading to a comparable increase in the intensity of EPEs in the 21st century (Westra et al., 2014; Sillmann et al., 2013). The region therefore anticipates more severe floods in future on account of the rise in both global and regional temperatures. Fischer et al. (2012) documented valuable information for stakeholders and decision-makers involved in coping strategies. It is not possible to evaluate the consequences of EPE accumulation and intensity directly from GCM or regional climate model (RCM)projections alone, since flood vulnerability depends not only on frequency and intensity of rainfall events but on other geo-environmental traits of topography, soil properties, human interventions in the form of physical flood protection measures, warnings, evacuation procedures and so on (Hijioka et al., 2014).

When reviewed in the Indian context, we found increasing EPEs in the place of weaker rainfall events over the central Indian region (IPCC, 2013b). In spite of having wide inter-seasonal variability in rainfall, it is increasing by 10 mm decade^{-1} in EPEs. The rainfall extremes have been witnessed since the ratio of rainfall in wet and dry seasons increased during 1955–2005. The occurrences of EPEs are likely to increase in the SAR, including the Indian subcontinent (IPCC, 2013b; Hijioka et al., 2014). There may be some regional differentiation in EPE intensities, but there is no contradiction in rising trends among researchers across the region (Table 2.1) (Rajeevan et al., 2008; Ghosh et al., 2009; Guhathakurta et al., 2011). In a recent study, Pai and Sridhar (2015) documented significant increasing trends of EPEs over all the geographical regions (central India [CI], north-eastern India [NEI] and the western coast [WC]) of India using high-resolution (0.25° × 0.25°) daily gridded rainfall data set covering 110 years (1901–2010), though the frequency of high extreme precipitation events (HEPEs) and very high extreme precipitation events (VHEPEs) significantly increased during the second half of the 20th century (1956–2010). By analyzing CORDEX South Asia RCM data, Krishnan and Sanjay (2017) disclosed that 5-day precipitation (RX5day) and simple daily intensity index (SDII) both are projected to increase by 21% and 38%, respectively, under RCP 8.5 overlapped with RCP 4.5 in the 21st century. Notwithstanding, their concentrations are expected to occur along the WC and the adjoining peninsular region.

Extreme precipitation events and flood vulnerability in South Asian river basins

Flood vulnerability generally refers to the degree to which a river system is unable to cope with the adverse effect of EPEs, while flood risks include exposure and vulnerability (IPCC, 2012; UNISDR, 2011). The exposure, risks and dimension of flood hazards have recently been investigated globally by Jongman et al. (2015), Pappenberger et al. (2012), Ward et al. (2013) and Winsemius et al. (2013). The International Disaster Database

(EM-DAT) claimed nearly 125 riverine flood records year^{-1} during 2001–2015, along with the loss of >4,000 lives annually across the globe. Increasing flood risks due to EPEs have also been projected for different countries in the South Asian humid region through GCM studies (Hirabayashi et al., 2013; Dastagir, 2015; Krishnan and Sanjay, 2017; Alahacoon et al., 2018). Flood casualties would be the highest in India, Bangladesh and Pakistan (AR5), even in port cities of South Asia like Kolkata, Mumbai and Dhaka, which are exposed to coastal flooding that would bear the maximum risk by the 2070s (IPCC, 2014). The El Niño or La Niña phases of the El Niño and the Southern Oscillation (ENSO) cycle (Ward et al., 2013) have been established to have high correlation with flooding in the river basins of Bangladesh, India and Pakistan (Ward et al., 2013).

During the last thirty years of the present century (1976–2005), India has experienced the most flood events (40%), followed by Bangladesh (17.2%), Pakistan (12.3%), Afghanistan (12.0%), Sri Lanka (10.2%), Nepal (7.2%), Bhutan (0.9%) and the Maldives (0.3%) (Shrestha, 2008) (Table 2.3). Dhar and Nandargi (2003) reported that most Indian flood events were confined to the Indo-Gangetic plain (IGP), northeast India and occasionally the rivers of central India. Nearly 40 Mha (12% of the geographical area [GA]) was affected by floods annually, while in Bangladesh, about 80% of the GA was exposed to flood hazards (Shrestha, 2008). From a long-term (1915–2015) perspective, India experienced 302 (47%) multi-dimensional flood hazards out of 649 of all types of disasters. Some of the hardest hitting devastating floods in the last five years were the Leh-Ladakh flood (2010), Assam flood (2012, 2017), Uttarakhand flood (2013), Jammu-Kashmir flood (2014), Manipur flood (2015) and recently Kerala flood (2018). As a consequence, the decadal economic burden burgeoned to US$34.5 billion (2005–2015) from US$11.6 billion in 1995–2005 (Tripathi, 2015).

Bangladesh is one of the most climate-risk-bearing countries in the world. It ranked fifth in global climate risk index (GCRI) among the 170 most vulnerable countries to climate change and is the sixth most flood-prone country in the world (Dastagir, 2015). Because of its unique geographical location, the country occupies only 7% of the catchment area of the Ganges–Brahmaputra–Meghna river system but is bound to drain >90% of the flow into the Bay of Bengal. Under the ongoing warming scenarios, the glaciers are melting rapidly in the source region of the Himalaya–Tibetan Plateau, causing floods in the lower catchment by accumulating surplus discharge in these areas. In the past 40 years, the country has experienced seven (in 1987, 1988, 1998, 2004 and 2007) devastating floods (Ghatak et al., 2012; Dewan, 2015). About 21% of the GA is subject to annual flooding, while an additional 42% of the GA is at the risk of floods with varied intensities. Apart from global warming, strategic location, rapid urbanization and burgeoning population pressure make the country the most vulnerable to climate change and flooding.

Table 2.3 Occurrences of extreme climatic events in the past three decades over eight countries in the South Asia Region, along with their implications

Countries	*Decade*	*Droughts*					*Depressions and storms*					*Extreme heat waves*					*Floods*				
		1	2	3	4	5 (1000s US$)	1	2	3	4	5 (1000s US$)	1	2	3	4	5 (1000s US$)	1	2	3	4	5 (1000s US$)
Afghanistan	1990–1999						1	10	0	0	0	2	224	0	200	0	12	1990	300	182305	64000
	2000–2009	3	37	0	4760000	50	3	331	0	22656	5000	4	1665	182	370684	10	41	1175	267	159042	20000
	2010–2018	2	0	0	3950000	142000	4	151	15	9060	0	1	45	68	68	0	25	1199	590	285024	3000
Bangladesh	1990–1999						44	142594	198817	23657965	2707500	8	801	0	34000	0	28	2944	2070	58165663	5070300
	2000–2009	1	0	0	0	0	40	5588	73558	13282840	2570000	9	1335	2200	203200	0	22	2404	322	58649753	2814000
	2010–2018						20	498	1402	9143246	664000	4	168	0	177000	0	14	539	21098	22156407	978000
Bhutan	1990–1999						1	17	0	65000	0						1	22	0	600	0
	2000–2009						1	12	0	0	0						2	200	0	1000	0
	2010–2018																				
India	1990–1999	2	0	0	1175000	542400	26	18091	8405	32346701	8029500	8	4039	0	0	0	44	13231	130	2.79E+08	9451982
	2000–2009	3	20	0	3.5E+08	1498722	30	1008	5339	5931589	369416	16	4926	50	50	400000	97	13666	416	2.47E+08	16800347
	2010–2018	1	0	0	3.3E+08	3000000	45	2555	1000	15839700	11584096	14	4419	0	0	0	69	11770	5804	76620022	32635500
Maldives	1990–1999						1	0	0	23849	30000										
	2000–2009																1			1649	
	2010–2018																				
Nepal	1990–1999						2	26	19	184	0						9	2537	500	809700	230200
	2000–2009	2	0	0	503000	0						3	126	200	200	0	11	1065	482	2089434	68729
	2010–2018						1	83	175	175	0	4	141	0	25000	123	15	1204	429	1942088	611000
Pakistan	1990–1999	1	143	0	2200000	247000	8	956	241	536371	10936	5	684	0	250	0	14	4180	115	18148606	1092230
	2000–2009						5	369	69	1653069	1620000	7	527	324	324	0	11	1065	482	2089434	68729
	2010–2018						5	99	386	4386	80000	3	1548	0	80000	18000	27	4639	9771	36494539	18113000
Sri Lanka	1990–1999																16	105	0	3325736	283880
	2000–2009	1	0	0	1000000	0	2	14	0	425000	0						17	363	2	3836067	35050
	2010–2018	3	0	0	4500000	45000	4	56	118	271457	403000						18	905	289	5239952	2199200

[1]Frequency, [2]Total deaths, [3]Injured, [4]Total affected and [5]Total damages.
Source: EM-DAT 2016

Increasing trends of EPEs (>100 mm within 24 hr occurred in the years of 2011, 2012 and 2016) are also reported from Sri Lanka (Alahacoon et al., 2018). Similarly, the RCMs projected consistent precipitation increases for all three scenarios (A2, A1B and B1) of 3.6–11.0%, 15.8–25% and 31.3–39.6% by 2030, 2050 and 2080, respectively, for Sri Lanka (Turner and Annamalai, 2012). The historical record (1966–2016) also showed a general increase of major flood events in Sri Lanka (Table 2.3) (EM-DAT, 2016).

Increase in temperature, EPEs, rapid snow melt, landslides, floods and glacial lake outburst floods make Nepal the fourth most vulnerable (among 170 countries) countries in the world (CCNN, 2011; MoSTE, 2015b; Anup, 2018). The country witnessed major floods in Tinao basin (1978), Kosi River (1980), Tadi River Basin (1985) and Sunkoshi Basin (1987) and a devastating cloud burst in the Kulekhani area (1993), which alone claimed 1336 lives (Ghatak et al., 2012). People of this high mountainous country are highly dependent on natural resources and agriculture; hence, slight changes in normal climate patterns make people more vulnerable (GWP Nepal, 2014; Anup, 2018). Climate change–induced floods also have been reported from Bhutan and Pakistan (Ali et al., 2017) (Table 2.3).

Occurrences and risks of cyclonic events, their frequencies and intensities

The North Indian Ocean (NIO) in the extreme south of the region is recognized as one of the breeding grounds of a majority of destructive tropical cyclones (TCs). TCs in general originate over the Bay of Bengal and the Arabian Sea and thereafter move in the north-northwest and north-east directions and finally strike on the coast of the SAR, especially the coasts of India, Bangladesh, Sri Lanka and Pakistan (Figure 2.2).

The landfall of these TCs in association with gusty wind, torrential precipitation and high storm surge create huge damage to infrastructure and loss of properties and life. Recent occurrences of high intensive TCs like Phailin (2013), Hudhud and Nilofar (2014), Luban and Titli (2018) and so on over the NIO draw the attention of regional as well as global policymakers and planners, climatologists, government bodies and other researchers towards varied sectors for their enormous impacts. The global warming scenarios stipulate more attention in view of their frequent origin, severe intensities and future risks. Table 2.3 shows the occurrences and vulnerabilities of TCs across different coastal areas of the SAR region (as published by EM-DAT) during the past three decades (1990–2018). Studies by Emanuel (2005) and Webster et al. (2005) revealed an increasing trend of frequency and intensity (power dissipation index; PDI) of TCs across nine basins of the world. Corroborating those findings, Singh et al. (2000) also reported increasing occurrences of TCs in the months of November and

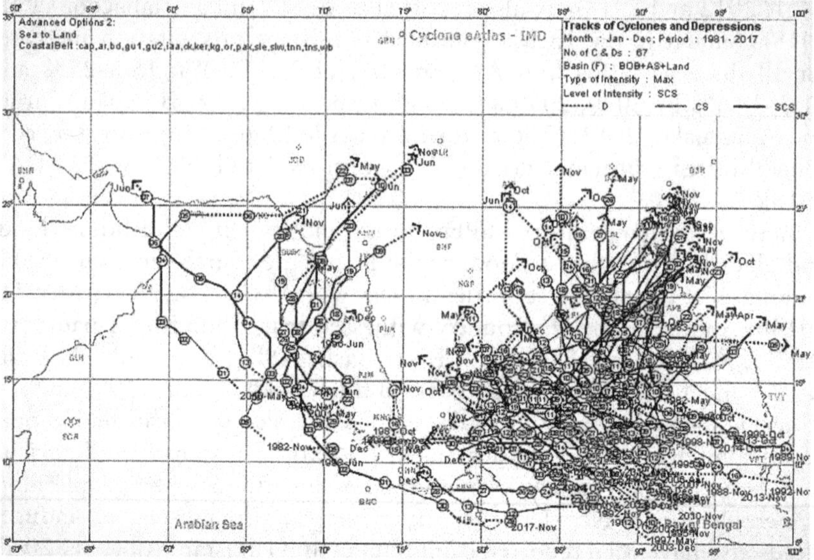

Figure 2.2 Tracks of severe cyclonic storms across the North Indian Ocean, Bay of Bengal and Arabian Sea for the period 1981–2017

Source: Cyclone e Atlas. www.rmcchennaieatlas.tn.nic.in (Retrieved on 18 January 2018)

May over the BOB. Although the overall frequency of monsoon depression showed a declining trend in the recent past, the transitional speed of TCs has increased (Singh et al., 2000; Srivastava et al., 2000; Niyas et al., 2009), along with an increase in the transitional speed of the depression in the BOB with a simultaneous decrease of that in the AS (Geetha and Balachandran, 2014). Apart from these, analysis of inter-annual fluctuation has also suggested increasing intensities of TCs in the recent past (Hoarau et al., 2012; Deo and Garner, 2014; Patra, 2015; Singh et al., 2018). In the 5th Assessment Report (AR5), the IPCC confirmed (though with low confidence for lack of systematic data) an increase in the number of high intensive TCs during the pre-satellite era (1980). However, it projected increasing intensities over some of the basins and decreasing one over the NIO but that precipitation from TCs would increase in future (Christensen et al., 2013). Also, the sea surface temperature had been significantly correlated with TC energy in the BOB and the El Niño or La Niña cycles, and the Indian Ocean dipole mode index (DMI) had high correlation with TCs during the post-monsoon season (Johnson and Xie, 2010; Ng and Chan, 2011; Evans and Waters, 2012). Most of the CMIP5 model studies revealed that the frequency of TCs will decrease, while the intensity of TCs will increase under the different RCP scenarios over the NIO region (Murakami et al., 2013; Yoshida et al.,

2017). A study by Emanuel (2013), on the contrary, established an accelerating frequency of TCs in the near (2030–2040) and distant (2100) futures under RCP 8.5 scenarios. Again, Walsh et al. (2015) reported increasing intensity but decreasing frequency of TCs across the SAR. Murakami et al. (2013) documented non-significant change in TC frequency over the NIO, with a substantial increase (46%) in the AS and corresponding decrease (31%) in the BOB. Not only that, the seasonal (pre-monsoon to monsoon) and spatial (westward) shifting of TCs over the NIO also has been reported (Murakami et al., 2013) (Figure 2.2).

Need for adaptation

If the region does not follow the IPCC recommendations for climate change mitigation and does not change its use of fossil fuels to power the SAR economy, then India, Bangladesh, Nepal, Sri Lanka, Bhutan and Maldives could lose nearly 1.8% of their gross domestic product (GDP) by 2050 and almost 9% by 2100 (ADB, 2014; Brears, 2015). Owing to these predictions, the developing economies of the SAR with relatively low preparedness could increasingly be affected by more frequent and intense ECEs (Shaw et al., 2016). Global climate change is manifested by its various direct and indirect effects in the SAR; among them, SMW anomaly-related heat waves, prevailing droughts, floods and TCs are gaining importance. Hence, it is urgent primarily to avoid heating processes and in turn reduce the inevitable disasters that are closely linked to climatic alterations in the SAR (World Bank, 2012). Many of those impacts in the region appear quite severe even with relatively modest warming of 1.5–2°C and pose significant challenges to all-round development. Thus, major coping options should be developed to sustain both regional economic prosperity and environmental sustainability.

Climate change hazards and development of resilience

Building resilience into the impacts of climate change requires identifying the risks and vulnerabilities in each event and sector. Identifying the feasible options for adaptation and mitigation measures is another important aspect of economic viability for all-round development of the community. Successful adaptation depends on sustained engagement of the stakeholders, inclusion of local to global organizations, public and private partnerships and, above all, local cooperation. In the SAR, not enough attention has been given to strengthen the lowest level of the governing structures, like Panchayats (village councils) and urban local bodies (ULBs), though these can play a significant role in this regard (Bhatt et al., 2019). Besides, many coping measures can be identified to address the issues, but prioritization is a critical aspect to adopt (Ahmed and Suphachalasai, 2014). Location-specific disaster preparedness must be composed of indigenous coping strategies, early warning systems and adaptive measures. Such integrated processes, however, require changes in the way the governments deal with policy-making, budgeting, implementation and monitoring at national, sector and project levels. Finally, adequate funding and technology transfer are important in ensuring success of mitigation options. The next section of this chapter deals with some of the pressing issues of climatic events in the SAR that have been occurring in the recent past and consequently the adaptation policies that should follow for the security of human life and property.

Reducing risks of flood hazards

Measures to manage floods may consist of local flood prevention schemes involving concrete structures and other engineered defenses such as dams, dikes and weirs. The loss of flood plains takes away key ecosystem services, including loss of biodiversity, water retention surfaces and prevention of soil erosion. Therefore, maintaining intact floodplains as part of an integrated flood management (IFM) system plays an important role in alleviating floods by storing water and releasing it slowly back into streams and rivers. Human-made flood defenses can also increase the vulnerability of communities, as has been seen in the Ganga-Brahmaputra delta region (Chapman and Rudra, 2007; Brears, 2015).

Approaches of drought hazard management

The global database of disasters reveals that drought constitutes nearly 3.5% among all natural disasters in Asia, whereas >80% of drought-affected people around the globe are from Asia, mostly because of weak economic strength vis-à-vis low coping options. The direct linkage between droughts and precipitation anomalies can be reduced over time with proper surface and subsurface water management systems. Adoption of appropriate tillage practices and planting more drought-resistant crop varieties can diminish the effect of drought significantly by conserving soil water and reducing transpiration (Mandal et al., 2015). In the drought-prone areas of Pakistan, more specifically in Baluchistan, traditional systems of irrigation have been developed. In Quetta valley and parts of Makran, where groundwater is accessible, *karez* are dug from springs to lead the water to the valley floor. Long tunnels are dug, with the help of shafts at 50-m intervals, to tap groundwater and maximize its flow to fields, gardens and orchards (SAARC Disaster Management Centre, 2010).

Coping options against tropical cyclones

TCs of differing intensity are active in major parts of the tropical and subtropical coastal regions, and these severely affect developing and LDCs, distressing nearly 3 million people year^{-1} in the SAR. In 2010, human vulnerability to TCs in LDCs was about <20% compared to that in 1980, though it was still 225 times higher than that in Organization for Economic Cooperation and Development (OECD) countries. Although human vulnerability to TCs has decreased to some extent, economic losses associated with TCs have increased. The average annual cost to GDP due to exposure to TCs from observed events in the SAR is approximately US$4.3 billion, a 14-fold increase from that in 1970 (World Bank, 2012). Implementation of early warning systems, peoples' awareness, risk-prone area identification and coastal area management should play a vital role in reducing vulnerabilities vis-à-vis risks of TCs in the region.

Adaptive and developmental approaches

A greater role of local communities in decision-making (Alauddin and Quiggin, 2008) and in prioritization and adoption of adaptation options (Prabhakar et al., 2010; Prabhakar and Srinivasan, 2011) is prescribed for better adaptation to adverse effects of climate change in the SAR. Defining adequate community property rights, reducing income disparity, exploring market-based and off-farm livelihood options, moving from production-based approaches to productivity- and efficiency-based approaches and promoting integrated decision-making approaches have also been suggested by researchers worldwide (Paul et al., 2009; Stucki and Smith, 2011). Climate-resilient livelihoods can be fostered through the creation of bundles of all natural, physical, human, financial and social capital and poverty eradication measures. Greater emphasis on agricultural growth has been suggested as an effective means of reducing rural poverty (Janvry and Sadoulet, 2010; Rosegrant, 2011). Community-based approaches may be suggested to identify adaptation options that address poverty and livelihoods (Huq and Reid, 2007) and help integration of disaster risk reduction, development and climate change adaptation (Heltberg et al., 2010), as well as to address location-specific adaptation (Rosegrant, 2011). Some occupational groups in India can become more vulnerable to change after being 'locked into' specialized livelihood patterns (Coulthard, 2008). Livelihood diversification is recommended as an important adaptation option for buffering climate change impacts on certain kinds of livelihoods (Keskinen et al., 2010). Ecosystem-based livelihood adaptations have always been found to show promising results, and integrating the use of biodiversity and ecosystem services can be an effective strategy to help people-centric development at the grassroots level (CDKN, 2014).

Concluding remarks and the way forward

Uncertainties in prediction models vis-à-vis a lack of uniform historical data across the SAR region is the greatest challenge for proper anticipation and exact identification of its ECE hotspot areas. More rigorous research on ECEs is therefore needed to understand their dimensions and to identify the exact points of locus much before their onset so that accurate location-specific adaptation and mitigation options may be adopted. Moreover, the weak economies of most of the South Asian countries are hindrances to establishing sophisticated laboratories and expertise in developing technical skills and implementing superior technology – like analysis of big data, artificial intelligence and robotics – for early warning of such stochastic processes of weather and climate. Development of local-level adaptation and mitigation strategies is imperative to reduce undesirable human interventions on natural ecosystems. The concerned decision-making bodies must be aware that the deployed mitigation options in each sector have both potential positive effects (synergies) and negative effects (trade-offs) in the domain of Sustainable Development Goals (SDGs) for each of the countries. Also, it is essential to integrate and cooperate in areas like sharing knowledge and experiences and the exchange of academic strengths to encourage joint research projects in this regard so that trans-boundary issues as well as regional warming scenarios as a whole may shrink to a desirable level. In a nutshell, the present chapter has drawn a concrete scenario of present and future vulnerabilities and best possible adaptive options for the occurrences of ECEs across the SAR and thus can contribute significantly to the growing body of scientific literature on climate change in the region. Given the diversified geographical setting, size and nature of the population and geopolitical significance of the region, it is important that countries cooperate with each other to achieve the common goals of economic and social prosperity through reducing the damages, including those from ECEs, which have been increasing in their frequencies and magnitudes during the recent past. The South Asian Association for Regional Cooperation, along with other organizations, should be reinvigorated to serve this purpose.

References

Ahmed, M., and Suphachalasai, S. 'Assessing the Cost of Climate Change and Adaptation in South Asia. Manila: ADB. 21st Century.' *Science*, 305 (2014): 994–997.

Aich, V., Akhundzadah, N. A., Knuerr, A., Khoshbeen, A. J., Hattermann, F., Paeth, H., Scanlon, A., and Paton E. N. 'Climate Change in Afghanistan Deduced from Reanalysis and Coordinated Regional Climate Downscaling Experiment (CORDEX) – South Asia Simulations.' *Climate*, 5(38) (2017). https://doi.org/10.3390/cli5020038

Ajayamohan, R. S., and Rao, S.A. 'Indian Ocean Dipole Modulates the Number of Extreme Rainfall Events over India in a Warming Environment.' *Journal of the Meteorological Society of Japan* (2008). https://doi.org/10.2151/jmsj.86.245

Akram, N., and Hamid, A. 'Climate Change: A Threat to the Economic Growth of Pakistan.' *Progress in Development Studies*, 15(1) (2015): 73–86.

Alahacoon, N., Matheswaran, K., Pani, P., and Amarnath, G.A. 'Decadal Historical Satellite Data and Rainfall Trend Analysis (2001–2016) for Flood Hazard Mapping in Sri Lanka.' *Remote Sensing*, 10 (2018): 448. https://doi.org/10.3390/rs10030448.

Alauddin, M., and Quiggin, J. 'Agricultural Intensification, Irrigation and the Environment in South Asia: Issues and Policy Options.' *Ecological Economics*, 65(1) (2008): 111–124.

Alexander, L., and Arblaster, J. M. 'Historical and Projected Trends in Temperature and Precipitation Extremes in Australia in Observations and CMIP5.' *Weather and Climate Extremes*, 15 (2017): 34–56.

Ali, S., Liu, Y., Ishaq, M., Shah, T., Ilyas, A. A., and Uddin, Z. 'Climate Change and Its Impact on the Yield of Major Food Crops: Evidence from Pakistan.' *Foods*, 6(39) (2017): 1–19. https://doi.org/10.3390/foods6060039. Available at: www.mdpi.com/journal/foods

Anjum, B.F., Khan, A.H., and Mir, H. 'Climate Change Perspective in Pakistan.' *Pakistan Journal of Meteorology*, 2(2) (2005): 11–21.

Anup, K. C. 'Climate Change Communication in Nepal.' In W. Leal Filho et al. (Eds.), *Handbook of Climate Change Communication* (Vol. II, pp. 21–35). 2018. https://doi.org/10.1007/978-3-319-70066-3_2

Ashfaq, M., Shi, Y., Tung, W.W., Trapp, R.J., Gao, X., Pal, J.S., and Diffenbaugh, N.S. 'Suppression of South Asian Summer Monsoon Precipitation in the 21st Century.' *Geophysical Research Letters*, 36 (2009): L01704. https://doi.org/10.1029/2008GL036500.

Asian Development Bank (ADB). 'Assessing the Costs of Climate Change and Adaptation in South Asia.' 2014. Available at: www.adb.org/publications/assessing-costs-climate-change-and-adaptation-south-asia (Accessed on 10 February 2015).

Asian Development Bank (ADB). 'A Region at Risk the Human Dimensions of Climate Change in Asia and the Pacific.' Manila, Philippines, 2017. http://dx.doi.org/10.22617/TCS178839-2

Bajracharya, A. R., Bajracharya, S. R., Shrestha, A. B., and Maharjan, S. B. 'Climate Change Impact Assessment on the Hydrological Regime of the Kaligandaki Basin, Nepal.' *Science of the Total Environment*, 625 (2018, June 1): 837–848.

Bal, P. K., Ramachandran, A., Palanivelu, K., Thirumurugan, P., Geetha, R., and Bhaskaran, B. 'Climate Change Projections over India by a Downscaling Approach Using PRÉCIS.' *Asia-Pacific Journal of Atmospheric Sciences*, 42(4) (2016, August).

Bandyopadhyay, N., Bhuiyan, C., and Saha, A. K. 'Heat Waves, Temperature Extremes and Their Impacts on Monsoon Rainfall and Meteorological Drought in Gujarat, India.' *National Hazards*, 82(2016): 367–388. https://doi.org/10.1007/s11069-016-2205-4

Baynes, C. 'Worst Floods in Nearly a Century Kill 44 in India's Kerala State amid Torrential Monsoon Rains.' *The Independent*, 2018. Available at: https://www.independent.co.uk/news/world/asia/india-worst-floods-flooding-death-monsoon-rain-dead-kerala-kochi-a8493011.html (Accessed on 16 August 2018).

Bhatt, M, Gleason, M. K., and Patel, R. B. 'Natural Hazards Governance in South Asia. 2019.' https://doi.org/101093/acrefore/9780199389407.013.231. Available at: http://oxfordre.com/naturalhazardscience/view/10.1093/acrefore/9780199389407.001.0001/acrefore9780199389407-e-231 (Accessed on 15 February 2019).

Bhutan Climate Change Handbook. Thimpu, Bhutan: Bhutan Media and Communications Institute, 2016.

Brears, R. 'Integrated Flood Risk Management in India and the South Asia Region: Lessons from the Rhine and Danube.' NFG Policy Paper Series, No. 08,

February 2015, NFG Research Group, Asian Perceptions of the EU Freie Universität, Berlin.

Chandrapala, L. 'Long Term Trends of Rainfall and Temperature in Sri Lanka.' In Y.P. Abrol, S. Gadgil, and G.B. Pant (Eds.), *Climate Variability and Agriculture* (pp. 153–162). New Delhi: Narosa Publishing House, 1996.

Chapman, G.P., and Rudra, K. 'Water as Foe, Water as Friend: Lessons from Bengal's Millennium Flood.' *Journal of South Asian Development,* 2(1) (2007). Sage Publications. https://doi.org/10.1177/097317410600200102

Chaudhry, Q.Z., Arif, M., Rasul, G., and Afzaal, M. *Climate Change Indicators of Pakistan.* Pakistan Meteorological Department Technical Report No. PMD-22/2009: 4–5 Islamabad, Pakistan, 2009.

Christensen, J. H., Hewitson, B., Busuioc, A., Chen, A., Gao, X., Held, I., Jones, R., Kolli, R. K., Kwon, W-T., Laprise, R., Rueda, V. M., Mearns, L. O., Menendez, C. G., Räisänen, J., Rinke, A., Sarr, A., and Whetton, P. 'Regional Climate Projections.' In *Climate Change 2007: The Physical Science Basis. Contribution of Working Group I to the Fourth Assessment Report of the Intergovernmental Panel on Climate Change* (pp. 847–940). Cambridge, UK and New York: Cambridge University Press, 2007.

Christensen, J. H., Krishna Kumar, K., Aldrian, E., An, S.-I., Cavalcanti, I.F.A., de Castro, M. Dong, W., Goswami, P., Hall, A., Kanyanga, J.K., Kitoh, A., Kossin, J., Lau, N.-C., Renwick, J., Stephenson, D.B., Xie, S.-P., and Zhou, T. 'Climate Phenomena and Their Relevance for Future Regional Climate Change.' In T.F. Stocker, D. Qin, G.-K. Plattner, M. Tignor, S. K. Allen, J. Boschung, A. Nauels, Y. Xia, V. Bex, and P. M. Midgley (Eds.), *Climate Change 2013: The Physical Science Basis. Contribution of Working Group I to the Fifth Assessment Report of the Intergovernmental Panel on Climate Change* (pp. 1–92). Cambridge, UK and New York: Cambridge University Press, 2013.

Climate and Development Knowledge Network (CDKN) and Overseas Development Institute (ODI). *The IPCC's Fifth Assessment Report. What's in It for South Asia? A Guide to the IPCC's Fifth Assessment Report Prepared for Decision-Makers in South Asia* (p. 187). Cambridge, UK and New York: Cambridge University Press, 2014.

Climate Change Network Nepal (CCNN). *Governance of Climate Change Adaptation Finance in Nepal.* Climate Change Network, Kathmandu, Nepal, 2011.

Cline, W. R. *Global Warming and Agriculture: Impact Estimates by Country.* Washington, DC: Peterson Institute for International Economics, 2007.

Coulthard, S. 'Adapting to Environmental Change in Artisanal Fisheries – Insights from a South Indian Lagoon.' *Global Environmental Change,* 18(3) (2008): 479–489.

Cruz, R.V., Harasawa, H., Lal, M., Wu, S., Anokhin, Y., Punsalmaa, B., Honda, Y., Jafari, M. Li, C., and HuuNinh, N. 'Asia. Climate Change 2007: Impacts, Adaptation and Vulnerability.' In M. L. Parry, O. F. Canziani, J. P. Palutikof, P.J. van der Linden, and C.E. Hanson (Eds.), *Contribution of Working Group II to the Fourth Assessment Report of the Intergovernmental Panel on Climate Change* (pp. 469–506). Cambridge, UK: Cambridge University Press, 2007.

Dai, A, Trenberth, K.E., and Qian, T. 'A Global Data Set of Palmer Drought Severity Index for 1870–2002: Relationship with Soil Moisture and Effects of Surface Warming.' *Journal of Hydrometeorology,* 5(2004): 1117–1130.

Dastagir, M. R. 'Modelling Recent Climate Change Induced Extreme Events in Bangladesh: A Review.' *Weather and Climate Extremes*, 7(2015): 49–60.

De Costa, W. A. J. M. 'Climate Change in Sri Lanka: Myth or Reality? Evidence from Long-Term Meteorological Data.' *Journal of the National Science Foundation of Sri Lanka*, 36 (2008): 63–88.

Deka, R. L., Mahanta, C., Pathak, H., Nath, K. K., and Das, S. 'Trends and Fluctuations of Rainfall Regime in the Brahmaputra and Barak Basins of Assam, India.' *Theory and Applied Climatology*, 114(1–2) (2012): 61–71. https://doi.org/10.1007/s00704-012-0820-x.

Deo, A. A., and Garner, D. W. 'Tropical Cyclone Activity over the Indian Ocean in the Warmer Climate, Change.' In U. C. Mohanty et al. (Eds.), *Monitoring and Prediction of Tropical Cyclone in the Indian Ocean and Climate* (pp. 72–80). W. M. Springer Science and Business Media, New Delhi, 2014.

De Silva, C. S., Weatherhead, E. K., Knox, J. W., and Rodriguez-Diaz, J. A. 'Predicting the Impacts Of Climate Change – A Case Study of Paddy Irrigation Water Requirements in Sri Lanka.' *Agricultural Water Management*, 93(2007): 19–29. http://dx.doi.org/10.1016/j.agwat.2007.06.003.

De Silva, G. J., and Sonnadara, D. U. J. 'Climate Change in the Hill Country of Sri Lanka.' Proceedings of the Technical Sessions, Institute of Physics, Sri Lanka, 2009.

Dewan, T. H. 'Societal Impacts and Vulnerability to Floods in Bangladesh and Nepal.' *Weather and Climate Extremes*, 7(2015): 36–42.

Dhar, O.N., and Nandargi, S. 'Hydrometeorological Aspects of Floods in India.' *Natural Hazards*, 28(2003): 1–33.

D. H. M. *Observed Climate Trend Analysis in the Districts and Physiographic Regions of Nepal (1971–2014)*. Kathmandu: Department of Hydrology and Meteorology, 2017.

Diffenbaugh, N.S., and Scherer, M. 'Observational and Model Evidence of Global Emergence of Permanent, Unprecedented Heat in the 20th and 21st Centuries.' *Climate Change*, 107(2011): 615–624.

Doosti, A. A., and Sherzad, M. H. *Climate Change and Governance in Afghanistan*. Kabul: National Environmental Protection Agency and United Nations Environment Programme, 2015.

Emanuel, K. A. 'Increasing Destructiveness of Tropical Cyclones over the Past 30 Years.' *Nature*, 436 (2005). https://doi.org/10.1038/nature03906.

Emanuel, K. A. 'Downscaling CMIP5 Climate Models Shows Increased Tropical Cyclone Activity over the 21st Century.' *PNAS*, 110(30) (2013): 12219–12224.

'EM-DAT, 2016: The Emergency Events Database – Université Catholique de Louvain (UCL) – CRED.' Available at: www.emdat.be, Brussels, Belgium (Accessed on 15 January 2016).

Eriyagama, N., Smakhtin, V., Chandrapala, L., and Fernando, K. *Impacts of Climate Change on Water Resources and Agriculture in Sri Lanka: A Review and Preliminary Vulnerability Mapping*. Colombo, Sri Lanka: International Water Management Institute, 2010.

Esham, M., and Garforth, C. 'Climate Change and Agricultural Adaptation in Sri Lanka: A Review.' *Climate and Development*, 5(1) (2013): 6676. https://doi.org/10.1080/17565529.2012.762333. Available at: http://centaur.reading.ac.uk/30166/

Evans, J. L., and Waters, J. J. 'Simulated Relationships between Sea Surface Temperatures and Tropical Convection in Climate Models and Their Implications for Tropical Cyclone Activity.' *Journal of Climate*, 25(2012): 7884–7895.

Fischer, J., Dyball, R., Fazey, I., Gross, C., Dovers, S., Ehrlich, R., Borden. R. J. 'Human Behaviour and Sustainability.' *Frontiers in Ecology and Environment*, 10(2012): 153–160, 10.1890/110079

Fischer, E. M., and Knutti, R. 'Anthropogenic Contribution to Global Occurrence of Heavy Precipitation and High-Temperature Extremes.' *Nature Climate Change*, 5(2015): 560–564.

Gautam, P. K. 'Climate Change and Conflict in South Asia.' *Strategic Analysis*, 36(1) (2012): 32–40. http://dx.doi.org/10.1080/09700161.2012.628482.

Gautam, R., Hsu, N. C., Lau, K.M., and Kafatos, M. 'Aerosol and Rainfall Variability over the Indian Monsoon Region: Distributions, Trends and Coupling.' *Annales Geophysicae*, 27(2009): 3691–3703. https://doi.org/10.5194/angeo-27-3691-2009.

Geetha, B., and Balachandran, S. 'Decadal Variations in Translational Speed of Cyclonic Disturbances over North Indian Ocean.' *Mausam*, 65(1) (2014): 115–118.

Ghatak, M., Kamal, A., and Mishra, O.P. (Back Ground Paper) 'Flood Risk Management in South Asia.' Proceedings of the SAARC Workshop on Flood Risk Management in South Asia, 2012.

Ghosh, S., Luniya, V., and Gupta, A. 'Trend Analysis of Indian Summer Monsoon Rainfall at Different Spatial Scales.' *Atmospheric Science Letters*, 10(2009): 285–290.

Ghulami, M. 'Assessment of Climate Change Impacts on Water Resources and Agriculture in Data-Scarce Kabul Basin, Afghanistan.' Other Université Côte d'Azur, 2017. English. <NNT: 2017AZUR4135>. <tel-01737052>.HAL Id: tel-01737052, 2018. Available at: https://tel.archives-ouvertes.fr/tel-01737052.

Goswami, B. N., Venugopal, V., Sengupta, D., Madhusoodanan, M.S., and Xavier, P.K. 'Increasing Trend of Extreme Rain Events over India in a Warming Environment.' *Science*, 314(2006): 1442–1445. https://doi.org/10.1126/science.1132027

Guhathakurta, P., Shreejith, O.P., and Menon, P.A. 'Impact of Climate Change on Extreme Rainfall Events and Flood Risk in India.' *Journal of Earth System Science*, 120(3) (2011): 359–373.

Gurung, G.B., and Bhandari, D. 'Integrated Approach to Climate Change Adaptation.' *Journal of Forest and Livelihood*, 8(1) (2009): 90–99.

Gütschow, J., Jeffery, M., Alexander, R., Hare, B., Schaeffer, M., Rocha, M., Höhne, N., Fekete, H., Breevoort, P. and Blok, K. 'INDCs lower projected warming to 2.7°C: Significant progress but still above 2°C.' Climate Action Tracker Update, 2015. Available at: http://climateactiontracker.org/publications/briefing/223/INDCs-lower-projected-warming-to-2.7C-significant-progress-but-still-above-2C-.html (Accessed on 15 February 2019).

GWP. 'Climate Change, Food and Water Security in South Asia: Critical Issues and Cooperative Strategies in an Age of Increased Risk and Uncertainty.' Synthesis of Workshop Discussions, a Global Water Partnership (GWP) and International Water Management Institute (IWMI) Workshop, 23–25 February 2011, Colombo, Sri Lanka Editors Tushaar Shah and Uma Lele, GWP Technical Committee Members.

GWP Nepal. *Climate Vulnerability and Gap Assessment Report on Flood and Drought (Lower Rapti River Basin Case Study)*. Kathmandu, Nepal: GWP Nepal/ Jalsrot Vikas Sanstha (JVS), 2014.

GWP Nepal. *Stocktaking: Climate Vulnerability on Agricultural Sector for National Adaptation Plan Process. Jalsrot Vikas Sanstha (JVS)*. Kathmandu, Nepal: GWP Nepal, 2015.

Handmer, J., Honda, Y., Kundzewicz, Z., Arnell, W.N., Benito, G., Hatfield, J., Mohamed, I.F., Peduzzi, P., Wu, S., Sherstyukov, B., Takahashi, K., and Yan, Z. 'Changes in Impacts of Climate Extremes: Human Systems and Ecosystems.' In C. B. Field, V. Barros, T. F. Stocker, D. Qin, D. J. Dokken, K. L. Ebi, M. D. Mastrandrea, K. J. Mach, G.-K. Plattner, S. K. Allen, M. Tignor, and P. M. Midgley (Eds.), *Managing the Risks of Extreme Events and Disasters to Advance Climate Change Adaptation*. A Special Report of Working Groups I and II of the Intergovernmental Panel on Climate Change (IPCC) (pp. 231–290). Cambridge, UK and New York: Cambridge University Press, 2012.

Hansen, J., Sato, M., and Ruedy, R. 'Perception of Climate Change.' *Proceedings of the National Academy of Science U.S.A.*, 109 (2012): E2415–E2423.

Heltberg, R., Prabhu, R., and Gitay, H. 'Community-Based Adaptation: Lessons from the Development Marketplace 2009 on Adaptation to Climate Change.' Social Development Working Papers, Paper No. 122, Social Development Department, The World Bank, Washington, DC, USA, June 2010, 53.

Herath, S., and Ratnayake, U.R. 'Monitoring Rainfall Trends to Predict Adverse Impacts A Case Study from Sri Lanka.' *Global Environmental Change*, 14(2004): 71–79. http://dx.doi.org/10.1016/j.gloenvcha.2003.11.009.

Hijioka, Y., Lin, E., Pereira, J.J., Corlett, R.T., Cui, X., Insarov, G.E., Lasco, R.D., Lindgren, E., and Surjan, A. 'Asia in Climate Change 2014: Impacts, Adaptation, and Vulnerability.' In V. R. Barros, C. B. Field, D. J. Dokken, M. D. Mastrandrea, K. J. Mach, T. E. Biller, M. Chattered, K. L. Ebi, Y. O. Estrada, R. C. Genova, B. Girma, E. S. Kissel, A. N. Levy, S. MacCracken, P. R. Mastrandrea, and L. L. White (Eds.), *Part B: Regional Aspects. Contribution of Working Group II to the Fifth Assessment Report of the Intergovernmental Panel on Climate Change* (pp. 1327–1370). Cambridge, UK and New York: Cambridge University Press, 2014.

Hirabayashi, Y., Mahendran, R., Koirala, S., Konoshima, L., Yamazaki, D., Watanabe, S., Kim, H., and Kanae, S. 'Global Flood Risk under Climate Change.' *Nature Climate Change*, 3(2013): 1–6. https://doi.org/10.1038/nclimate1911.

Hoarau, K., Bernard, J., and Chalonge, L. 'Review: Intense Tropical Cyclone Activities in the Northern Indian Ocean.' *International Journal of Climatology*, 32(2012): 1935–1945. Royal Meteorological Society, London. https://doi.org/10.1002/joc.5254.

Huq, S. R., and Reid, H. 'Community-Based Adaptation: A Vital Approach to the Threat Climate Change Poses to the Poor.' IIED Briefing, International Institute for Environment and Development (IIED), London, UK, 2, 2007.

Imbulana, K.A.U.S., Wijesekara, N.T.S., and Neupane, B.R. (Eds.). *Sri Lanka National Water Development Report*. Sri Lanka, Paris and New Delhi: MAI&MD, UN-WWAP, UNESCO and University of Moratuwa, 2006.

IPCC. 'Climate Change 2007: The Physical Science Basis.' In S. Solomon, D. Qin, M. Manning, Z. Chen, M. Marquis, K. B. Averyt, M. Tignor, and H. L. Miller (Eds.), *Contribution of Working Group I to the Fourth Assessment Report of the*

Intergovernmental Panel on Climate Change (p. 996). Cambridge, UK and New York: Cambridge University Press, 2007.

IPCC. 'Managing the Risks of Extreme Events and Disasters to Advance Climate Change Adaptation.' In C.B. Field, V. Barros, T.F. Stocker, D. Qin, D. J. Dokken, K. L. Ebi, M. D. Mastrandrea, K.J. Mach, G.-K. Plattner, S.K. Allen, M. Tignor, and P. M. Midgley (Eds.), *A Special Report of Working Groups I and II of the Intergovernmental Panel on Climate Change* (p. 582). Cambridge, UK and New York: Cambridge University Press, 2012.

IPCC. 'Climate Change 2013: The Physical Science Basis.' In T.F. Stocker, D. Qin, G.-K. Plattner, M. Tignor, S.K. Allen, J. Boschung, A. Nauels, Y. Xia, V. Bex, and P. M. Midgley (Eds.), *Contribution of Working Group I to the Fifth Assessment Report of the Intergovernmental Panel on Climate Change*. Cambridge, UK and New York: Cambridge University Press, 2013a. https://doi.org/10.1017/CBO9781107415324

IPCC. 'Summary for Policymakers. In Climate Change 2013: The Physical Science Basis.' In T.F. Stocker, D. Qin, G.-K. Plattner, M. Tignor, S.K. Allen, J. Boschung, A. Nauels, Y. Xia, V. Bex, and P. M. Midgley (Eds.), *Contribution of Working Group I to the Fifth Assessment Report of the Intergovernmental Panel on Climate Change*. Cambridge, UK and New York: Cambridge University Press, 2013b.

IPCC. 'Climate Change 2014: Synthesis Report.' In Core Writing Team, R. K. Pachauri, and L.A. Meyer (Eds.), *Contribution of Working Groups I, II and III to the Fifth Assessment Report of the Intergovernmental Panel on Climate Change* (p. 151). Geneva, Switzerland: IPCC, 2014.

Iqbal, W., and Zahid, M. 'Historical and Future Trends of Summer Mean Air Temperature over South Asia.' *Pakistan Journal of Meteorology*, 10(20) (2014): 67–74.

Janvry, A., and Sadoulet, E. 'Agricultural Growth and Poverty Reduction: Additional Evidence.' *The World Bank Research Observer*, 25(1) (2010): 1–20.

Jayatillake, H. M., Chandrapala, L., Basnayake, B. R. S. B., and Dharmaratne, G. H. P. 'Water Resources and Climate Change.' In N. T. S. Wijesekera, K. A. U. S. Imbulana, and B. Neupane (Eds.), *Proceedings of the Workshop on Sri Lanka*. Paris, France: National Water Development Report. World Water Assessment Programme (WWAP), 2005.

Johnson, N.C., and Xie, S.P. 'Changes in the Sea Surface Temperature Threshold for Tropical Convection.' *Nature Geoscience*, 3(2010): 842–845.

Jongman, B., Winsemius, H.C., Aerts, J.C.J.H., et al. 'Declining Vulnerability to River Floods and the Global Benefits of Adaptation.' *Proceedings of the National Academy of Sciences USA*, 112(2015): E2271–E2280. https://doi.org/10.1073/pnas.1414439112.

Keskinen, M. Chinvanno, S., Kummu, M., Nuorteva, P., Snidvongs, A., Varis, O., and Vastila, K. 'Climate Change and Water Resources in the Lower Mekong River Basin: Putting Adaptation into Context.' *Journal of Water and Climate Change*, 1(2) (2010): 103–117.

Kreft, S., and Eckstein, D. 'Who Suffers Most from Extreme Weather Events?' In *Global Climate Risk Index 2014*. Germanwatch, Germany, 2016.

Krishnan, R., and Sanjay, J. 'Climate Change over INDIA: An Interim Report.' ESSO-Indian Institute of Tropical Meteorology Ministry of Earth Sciences, Govt. of India. Centre for Climate Change Research, 1–38, 2017.

Kumar, K. R., Sahai, A. K., Krishna Kumar, K., Patwardhan, S. K., Mishra, P. K., Revadekar, J. V., Kamala, K., and Pant, G. B. 'High-Resolution Climate Change Scenarios for India for the 21st Century.' *Current Science*, 90(3) (2006): 334–345.

Mandal, S., and Choudhury, B.U. 'Estimation and Prediction of Maximum Daily Rainfall at Sagar Island Using Best Fit Probability Models.' *Theoretical and Applied Climatology*, 121(1–2) (2014): 87–97. https://doi.org/10.1007/s00704-014-1212.1.

Mandal, S., Choudhury, B.U., and Satpati, L. N. 'Monsoon Variability, Crop Water Requirement, and Crop Planning for Kharif Rice in Sagar Island, India.' *International Journal of Biometeorology* (2015). https://doi.org/10.1007/s00484-015-0995-9

Mandal, S., Satpati, L.N., Choudhury, B.U., and Sadhu, S. 'Climate Change Vulnerability to Agrarian Ecosystem of Small Island: Evidence from Sagar Island, India.' *Theoretical and Applied Climatology* (2017). https://doi.org/10.1007/s00704-017-2098-5

Mani, M., Bandyopadhyay, S., Chonabayashi, S., Markandya, A., and Mosier, T. 'South Asia's Hotspots: The Impact of Temperature and Precipitation Changes on Living Standards.' In *South Asia Development Matters*. Washington, DC: World Bank. https://doi.org/10.1596/978-1-4648-1155-5. License: Creative Commons Attribution CC BY 3.0 IGO, World Bank Publications, 2018, The World Bank Group, Washington, DC 20433, USA; ISBN (paper): 978-1-4648-1155-5, ISBN (electronic): 978–1 4648–1156–2, https://doi.org/10.1596/978-1-4648-1155-5.

Mazdiyasni, O., Agha Kouchak, A., Davis, S.J., Madadgar, S., Mehran, A., Ragno, E., Sadegh, M., Sengupta, A., Ghosh, S., Dhanya, C.T., and Niknejad, M. 'Increasing Probability of Mortality during Indian Heat Waves.' *Sciences Advances*, 3(6) (2017): e1700066.

Meteorological Department. *International Climate Technology Expert*. Asian Development Bank.

Mishra, V, Aadhar, S., Shah, H., Kumar, R., Pattanaik, D.R., and Tiwari, A.D. 'The Kerala Flood of 2018: Combined Impact of Extreme Rainfall and Reservoir Storage.' *Hydrology and Earth System Sciences Discussions* (2018). https://doi.org/10.5194/hess-2018-480

Miyan, M.A. 'Droughts in Asian Least Developed Countries: Vulnerability and Sustainability.' *Weather and Climate Extremes*, 7(2015): 8–23.

MOE. *National Adaptation Programme of Action to Climate Change*. Kathmandu, Nepal: Ministry of Environment, Government of Nepal, 2010.

MoSTE. *Indigenous and Local Climate Change Adaptation Practices in Nepal. Ministry of Science Technology and Environment*. Kathmandu, Nepal: Government of Nepal, 2015a.

MoSTE. *National Adaptation Plan Formulation Process*. Kathmandu, Nepal: Government of Nepal, Ministry of Science, Technology and Environment, Climate Change Management Division, 2015b.

Murakami, H., Sugi, M., Kitoh, A., et al. 'Future Changes in Tropical Cyclone Activity Projected by the New High Resolution MRI-AGCM.' *Journal of Climate*, 40 (2013): 1949–1968, https://doi.org/10.1007/s00382–012.1407-z.

National Adaptation Plan of Action. *Maldives: NAPA Draft for Comments Ministry of Environment, Energy and Water, Maldives*, 2006.

National Disaster Management Authority (NDMC). *Guidelines for Preparation of Action Plan Prevention and Management of Heat-Wave* (pp. 1–12). Government of India, New Delhi, 2016.

NCVST. *Vulnerability through the Eyes of Vulnerable: Climate Change Induced Uncertainties and Nepal's Development Predicaments*. Boulder, CO: Institute for Social and Environmental Transition-Nepal (ISET-N), Kathmandu and Institute for Social and Environmental Transition (ISET), 2009.

NEPA and UNEP. *Climate Change and Governance in Afghanistan* (pp. 1–36). Kabul: National Environmental Protection Agency and United Nations Environment Programme, 2015.

Ng, E.K.W., and Chan, J.C.L. 'Interannual Variations of Tropical Cyclone Activity over the North Indian Ocean.' *International Journal of Climatology*, 32(2011): 819–830. https://doi.org/10:1002.joc.2304.

Niyas, N.T., Srivastava, A.K., and Hatwar, H.R. 'Variability and Trend in the Cyclonic Storms over North Indian Ocean.' Meteorological Monograph No. Cyclone Warning – 3/2009, 2009, 35 pp. Available at: http://www.imdpune.gov.in/ncc_rept/Met_Monograph%20No.%203_2009.pdf.

Pai, D. S., and Sridhar, L. 'Long Term Trends in the Extreme Rainfall Events over India.' In K. Ray et al. (Eds.), *High-Impact Weather Events over the SAARC Region*. 2015. https://doi.org/10.1007/978-3-319-10217-7_15

Pappenberger, F., Stephens, E., Thielen, J., Salamon, P., Demeritt, D., Andel, S. J., Wetterhall, F., and Alfieri, L. 'Visualizing Probabilistic Flood Forecast Information: Expert Preferences and Perceptions of Best Practice in Uncertainty Communication.' *Hydrological Processes*, 27(2012): 132–146. https://doi.org/10.1002/hyp.9253.

Patra, P. 'Current Trends of Tropical Cyclone Energy: The North Indian Ocean (NIO) Perspective.' In V. P. Sati et al. (Eds.), *Climate Change and Socio-Ecological Transformation* (pp. 407–419). New Delhi: Today and Tomorrow's Printers and Publishers, 2015.

Paul, H., Ernsting, A., Semino, S., Gura, S., and Lorch, A. 'Agriculture and Climate Change: Real Problems, False Solutions.' Report prepared for the Conference of the Parties, COP15, of the United Nations Framework Convention on Climate Change in Copenhagen, December 2009 by Econexus, Biofuelwatch, Grupo de Reflexion Rural, NOAH – Friends of the Earth Denmark, and the Development Fund Norway, EcoNexus, Oxford, UK, 42 pp.

Prabhakar, S.V.R.K., Kobashi, T., and Srinivasan, A. (2010). 'Monitoring Progress of Adaptation to Climate Change: The Use of Adaptation Metrics.' *Asian Journal of Environment and Disaster Management*, 2(3) (2010): 435–442.

Prabhakar, S.V.R.K., and Srinivasan, A. 'Metrics for Mainstreaming Adaptation in Agriculture Sector.' In R. Lal, M. V. K. Sivakumar, S. M. A. Faiz, A. H. M. M. Rahman, and K. R. Islam (Eds.), *Climate Change and Food Security in South Asia* (pp. 551–568). Dordrecht, Netherlands: Springer Science, 2011.

Raihan, S. 'Next Steps to South Asian Economic Union: A Study on Regional Economic Integration (Phase II).' In S. Raihan (Ed.), *Prepared for South Asian Association for Regional Cooperation (SAARC)* (pp. 1–541). The South Asian Association for Regional Cooperation (SAARC) Secretariat, 2018.

Rahman, M. M., Islam, M. N., Ahmed, A. U., and Georgi, F. 'Rainfall and Temperature Scenarios for Bangladesh for the Middle of 21st Century Using Reg CM.' *Journal of Earth System Science* (2012): 287–295.

Rahman, M. R., and Lateh, H. 'Climate Change in Bangladesh: A Spatio-Temporal Analysis and Simulation of Recent Temperature and Rainfall Data Using GIS and Time Series Analysis Model.' *Theoretical and Applied Climatology* (2015): 1–15.

Rajbhandari, R., Shrestha, A. B., Nepal, S., and Wahid, S. 'Projection of Future Precipitation and Temperature Change over the Transboundary Koshi River Basin Using Regional Climate Model PRÉCIS.' *Atmospheric and Climate Sciences*, 8(2018): 163–191.

Rajeevan, M., Bhate, J., and Jaswal, A. K. 'Analysis of Variability and Trends of Extreme Rainfall Events over India Using 104 Years of Gridded Daily Rainfall Data.' *Geophysical Research Letters*, 35(2008): L18707. https://doi.org/10.1029/2008GL035143.

Rathore, L., Attri, S., and Jaswal, A. *State Level Climate Change Trends in India*. New Delhi: India Meteorological Department, 2013.

Rosegrant, M.W. 'Impacts of Climate Change on Food Security and Livelihoods.' In M. Solh and M. C. Saxena (Eds.), *Food Security and Climate Change in Dry Areas: Proceedings of an International Conference, 1–4 February 2010, Amman, Jordan* (pp. 24–26). Aleppo, Syria: International Centre for Agricultural Research in the Dry Areas (ICARDA), 2011.

Rupakumar, K., and Patil, S. D. 'Long-Term Variations of Rainfall and Surface Air Temperature over Sri Lanka.' In Y. P. Abrol, S. Gadgil, and G. B. Pant (Eds.), *Climate Variability and Agriculture*. New Delhi: Narosa Publishing House, 1996.

SAARC Disaster Management Centre. 'Drought Risk Management in South Asia.' Workshop Proceedings, New Delhi, 2010. Available at: http://droughtmanagement.info/literature/SAARC_drought_risk_management_south_asia_2010. pdf (Accessed on 15 February 2019).

Savage, M., Dougherty, B., Hamza, M., Butterfield, R., and Eharwani, S. *Socio-Economic Impacts of Climate Change in Afghanistan*. Report DFID CNTR 08 8507. Oxford: Stockholm Environmental Institute, 2009.

Schoof, J.T., and Robeson, S.M. 'Projecting Changes in Regional Temperature and Precipitation Extremes in the United States.' *Weather and Climate Extremes*, 11(2016): 28–40.

Shaw, R., Atta-ur-Rahman, Surjan, A., and Parvin, G. A. *Urban Disasters and Resilience in Asia*. Elsevier Inc., 2016. https://doi.org/10.1016/C2014-0-01952.1

Shi, L., Kloog, I., Zanobetti, A., Liu, P., and Schwartz, P.J.D. 'Impacts of Temperature and Its Variability on Mortality in New England.' *National Climate Change*, 5(2015): 988–991.

Shrestha, M.S. 'Impacts of Floods in South Asia.' *Journal of South Asia Disaster Studies*, 1(1) (2008): 85–106.

Sillmann, J, Kharin, V.V., Zwiers, F.W., Zhang, X., and Bronaugh, D. 'Climate Extremes Indices in the CMIP5 Multimodel Ensemble: Part 2. Future Climate Projections.' *Journal of Geophysical Research: Atmospheres*, 118(2013): 2473–2493. https://doi.org/10.1002/jgrd.50188.

Singh, K., Panda, J., Sahoo, M., and Mohapatra, M. 'Variability in Tropical Cyclone Climatology over North Indian Ocean during the Period 1891 to 2015.' *Asia-Pacific Journal of Atmospheric Sciences* (2018): 1–19. https://doi.org/10.1007/s13143-018-0069-0.

Singh, N, and Sontakke, N. A. 'On Climatic Fluctuations and Environmental Changes of the Indo-Gangetic Plains, India.' *Climate Change*, 52(2002): 287–313.

Singh, O.P., Khan, T.M.A., and Rahman, M.S. 'Changes in the Frequency of Tropical Cyclones over the North Indian Ocean.' *Meteorology and Atmospheric Physics*, 75(2000): 11–20.

Sivakumar, M.V.K., and Stefanski, R. 'Climate Change in South Asia.' In R. Lal et al. (Eds.), *Climate Change and Food Security in South Asia* (pp. 13–30). 2014. https://doi.org/10.1007/978-90-481-9516-9_2.

Srivastava, A.K., Sinham Ray, K.C., and De, U.S. 'Trends in the Frequency of Cyclonic Disturbances and Their Intensification over Indian Sea.' *Mausam*, 51(2000): 113–118.

State of Climate Change Report for the RNR Sector. *RNR Climate Change Adaptation Program* (pp. 7–33). Ministry of Agriculture & Forests, Royal Government of Bhutan, Bhutan, 2016.

Steffen, W., Hughes, L., and Perkins, S. *Heat Waves: Hotter, Longer, More Often.* Climate Council of Australia Limited, Sydney, 2014.

Stocker, T., Qin, D., Plattner, G.K., Tignor, M., Allen, S. K., Boschung, J., Nauels, A., Xia, Y., Bex, V., and Midgley, P.M. 'IPCC, Climate Change 2013: The Physical Science Basis.' In *Contribution of Working Group I to the Fifth Assessment Report of the Intergovernmental Panel on Climate Change* (p. 1535). Cambridge University Press, 2013.

Stucki, V., and Smith, M. 'Integrated Approaches to Natural Resources Management in Practice: The Catalyzing Role of National Adaptation Programmes for Action.' *AMBIO: A Journal of the Human Environment*, 40(4) (2011): 351–360.

Tollefson, J. 'Is the 2 °C world a fantasy?' *Nature*, 527 (26 November 2015): 436–438. 10.1038/527436a.

Trenberth, K.E., Fasullo, J.T., and Shepherd, T.G. 'Attribution of Climate Extreme Events.' *Nat. Climate Change*, 5(8) (2015): 725–730. https://doi.org/10.1038/Nclimate2657

Tripathi, P. 'Flood Disaster in India: An Analysis of Trend and Preparedness.' *Interdisciplinary Journal of Contemporary Research*, 2(4) (2015): 91–97.

Tse-ring, K. *Constructing Future Climate Scenario of Bhutan In Project Report on Climate Change Vulnerability and Adaptation Study for Rice Production in Bhutan for Project: Climate Change Studies in Bhutan, Activity No*. Thimpu: Ministry of Agriculture. Available at: www 094505-But.2.2003.

Tse-ring, K., Sharma, S., Chettri, N., and Shrestra, A. (Eds.). 'Climate Change Vulnerability of Mountain Ecosystems in the Eastern Himalayas.' In *Climate Change Impact and Vulnerability in the Eastern Himalayas*. Synthesis Report, ICIMOD, 2010.

Turner, A.G., and Annamalai, H. 'Climate Change and the South Asian Summer Monsoon.' *National Climate Change*, 2(2012): 587–595.

UNFCCC. *Climate Change: Impacts, Vulnerabilities and Adaptation in Developing Countries*. Bonn, Germany: Climate Change Secretariat (UNFCCC), 2007.

UNISDR. *Global Assessment Report on Disaster Risk Reduction*. Geneva, Switzerland: United Nations International Strategy for Disaster Reduction, 2011.

Vinke, K., Martin, M. A., Adams, S., Baarsch, F., Bondeau, A., Coumou, D., Donner, R. V., Menon, A., Perette, M., Rehfeld, K., Robinson, A., Rocha, M., Schaeffer, M., Schwan, S., Serdeczny, O., and Svirejeva-Hopkins, A. 'Climatic Risks and Impacts in South Asia: Extremes of Water Scarcity and Excess.' *Regional Environmental Change*, 17(6) (2017): 1569–1583.

Walsh, B. J., Rydzak, F., Palazzo, A. et al. 'New Feed Sources Key to Ambitious Climate Targets.' *Carbon Balance Manage* 10(26) (2015). https://doi.org/10.1186/s13021-015-0040-7

Ward, P. J., Pauw, W. P., van Buuren, M. W., and Marfai, M. A. 'Governance of Flood Risk Management in a Time of Climate Change: The Cases of Jakarta and Rotterdam.' *Environmental Politics*, 22(3) (2013): 518–536. https://doi.org/10.1080/09644016.2012.683155, 2013.

Webster, P.J., Holland, G.J., Curry, J.A., and Chang, H.R. 'Changes in Tropical Cyclone Number, Duration, and Intensity in a Warming Environment.' *Science (Reports)*, 309 (2005): 1844–1846. https://doi.org/10.1126/science.1116448.

Westra, S., et al. 'Future Changes to the Intensity and Frequency of Short-Duration Extreme Rainfall.' *Review of Geophysics*, 52(2014): 522–555.

Winsemius, H.C., Van Beek, L.P.H., Jongman, B., Ward, P.J., and Bouwman, A. 'A Framework for Global River Flood Risk Assessments.' *Hydrology and Earth System Sciences*, 17(2013): 1871–1892.

WMO. *Heat Waves and Health: Guidance on Warning-System Development*. Report No. 1142, Geneva, Switzerland, 2015.

Woo, S., Singh, G. P., Oh, J.-H., and Lee, K.-M. 'Projection of Seasonal Summer Precipitation over Indian Sub-Continent with a High-Resolution AGCM Based on the RCP Scenarios.' *Meteorology and Atmospheric Physics* (2018): 1–20.

World Bank. *Disaster Risk Management in South Asia: A Regional Overview*. Washington, DC: World Bank, 2012. Available at: https://openknowledge.worldbank.org/handle/10986/13218 (Accessed on 15 February 2019).

World Bank. 'Turn down the 4o Heat: Climate Extremes, Regional Impacts, and the Case for Resiliencies.' Executive Summary Report for the World Bank by the Potsdam Institute for Climate Impact Research and Climate Analytics, Washington, DC, 2013.

World Food Programme (WFP) and the UN Environment Programme (UNEP). *Climate Change in Afghanistan. What Does It Mean for Rural Livelihoods and Food Security?* (pp. 1–72). United Nations, Rome, 2016.

Yoshida, K., Sugi, M., Mizuta, R., Murakami, H., and Ishii, M. 'Future Changes in Tropical Cyclone Activity in High-Resolution Large-Ensemble Simulations.' *Geophysical Research Letters*, 44(2017): 9910–9917. https://doi.org/10.1002/2017GL075058.

Zahid, M., and Rasul, G. 'Changing Trends of Thermal Extremes in Pakistan.' *Climatic Change*, 113(3) (2012): 883–896.

Zhang, D.D., Brecke, P., Lee, H.F., He, Y.Q., and Zhang, J. 'Global Climate Change, War, and Population Decline in Recent Human History.' *Proceedings of the National Academy of Sciences*, 104(49) (2007): 19214–19219.

Zubair, L., Hansen, J., Yahiya, Z., Siriwardhana, M., Chandimala, J., Razick, S., Tennakoon, U., Ariyaratne, K., Bandara, I., Bulathsinhala, H., Abeyratne, T., and Samuel, T. D. M. A. 'Impact Assessment and Adaptation to Climate Change of Plantations in Sri Lanka.' IRI Technical Report, 2010.

3 Energy use and carbon dioxide emissions in South Asia

A decomposition analysis

Ujjaini Mukhopadhyay and Ratnakar Pani

The worldwide pursuit of economic growth and the consequential increased fossil fuel consumption has led to huge emissions of greenhouse gases, including carbon dioxide, resulting in global warming and climate change. In the recent years, the Asian countries have been the frontrunners in economic growth, with the rise in production being accompanied by escalating energy use, triggering rampant carbon dioxide emissions as well. According to World Bank data in 2014, South Asia accounts for 23% of the world population and contributes only 3% in global gross domestic product (GDP) but is responsible for 29% of the world's CO_2 emissions.

There exists considerable literature on the relationship between CO_2 emissions and their determinants. The studies can be viewed from four perspectives: the emission-population nexus, the emission-income nexus, the emission-technology nexus and the emission-population-income-technology nexus. The first group considers population growth a crucial factor in increasing emissions (Engleman, 1998; Crowley, 2000; Shi, 2001; Harte, 2007; York, 2007; Casey and Galor, 2016). The second group emphasizes the role of rising income in perpetuating emissions. While some studies find a positive monotonic relationship between income and emission levels (Shafik, 1994; Holtz-Eakin and Selden, 1995; Hang and Yuan-sheng, 2011), some focus on testing the environment Kuznets curve (EKC) hypothesis (Shafik and Bandyopadhyay, 1992; Coondoo and Dinda, 2002; Lipford and Yandle, 2010). The EKC hypothesis seeks to establish an inverted U-shaped relationship between income per capita and environmental degradation. It posits that at early stages of economic growth, environmental degradation rises at an increasing rate, but after some threshold level of economic growth, the movement tends to reverse (Grossman and Krueger, 1994). The third group proposes that energy and emission-saving technological innovation can be an effective means of carbon reduction (Hang and Yuan-sheng, 2011; Jiang and Hardee, 2011; Yii and Geetha, 2017). The fourth group, conducting decomposition study, considers that emission is determined collectively by a number of factors – population, income and technology being foremost among them.

Since emissions are the outcome of positive and negative impacts of various intertwined factors, each exerting its own push and pull effects, the decomposition approach has been the most popular one. Ang (2005) proposed the decomposition of change in emissions using the logarithmic mean divisia index (LMDI) method in order to determine the relative contribution of each of the factors in changing the emission level. The LMDI technique has been extensively used for decomposition analysis of emissions. Yeo et al. (2015) used LMDI to analyze the key drivers behind changes in CO_2 emissions in the residential sectors of China and India.

Mousavi et. al. (2017) performed three variations of decomposition analyses on driving forces of carbon emissions in Iran from 2003 to 2014 due to energy consumption of the industry, carbon intensity of electricity generation and fossil fuel combustion. Cansino et al. (2015) assessed the contribution of drivers of CO_2 emissions in Spain for the period 1995–2009. Wang et al. (2011) investigated the potential factors influencing the change of transport sector CO_2 emissions in China. Apart from country-level studies, few multi-country analyses have also been performed. Lee and Oh (2006) conducted a five-factor decomposition for 15 countries in the Asia Pacific Economic Cooperation (APEC) region between 1980 and 1998; World Bank (2007) which decomposed the CO_2 of the top 70 emitters between 1994 and 2004. Pani and Mukhopadhyay (2010) used panel data for a decomposition study of emissions in 114 countries. This chapter focuses on decomposition analysis of growing carbon dioxide emissions in South Asian countries.

South Asia has high population growth[1] and has been experiencing remarkable economic growth, particularly in the post-liberalization period (the growth rate of GDP per capita rising from –0.28% in 1991 to 5.61% in 2017, according to World Bank estimates). It has been increasingly concerned with arresting its escalating emissions level. However, the percentage of fossil fuel in total energy consumption has shot up remarkably, from 53.22% in 1991 to 71.25% in 2017. The region also attracts considerable inflow of foreign direct investment (FDI). As evident from World Bank estimates, net inflow of FDI has increased in South Asia from 0.11% of GDP in 1991 to 1.43% of GDP in 2017. FDI has two opposing effects on emissions: first, it raises total production, entailing higher fuel use, resulting in emissions, and may particularly encourage relocating polluting industries to developing countries with less stringent environmental regulations, thereby polluting the host countries more (pollution haven hypothesis).[2] Second, it involves transfer of advanced technology, including energy-efficient technology that reduces energy consumption, leading to lower relative emissions. It can be fairly presumed that CO_2 emissions in South Asia have been affected by several contributing factors.

Against the backdrop of the Paris Agreement on climate change signed by 195 countries in 2016, with commitments to reduce greenhouse gas emissions, it becomes imperative to objectively trace the perpetrating factors and find ways to arrest emissions in South Asia. This chapter uses the logarithmic mean divisia index method to undertake a decomposition analysis of the carbon dioxide emissions of six South Asian countries, Afghanistan, Bangladesh, India, Nepal, Pakistan and Sri Lanka,[3] over the period 1980–2015. The objectives of the study are first, to identify the role of the major contributing factors in changing emission levels; second, to ascertain whether FDI has any role in raising emissions; and third, to examine the extent to which changes in energy consumption pattern can regulate the emission level.

Data and methodology

Data source

The data source for carbon dioxide emissions, fossil fuel consumption, primary energy consumption and population is the US Energy Information Administration (EIA). The emission of CO_2 is measured in million metric tons; fossil fuel and primary energy consumption are given in terms of quadratic Btu (10^{12} British thermal units); population figures are given in millions. The data on GDP and investment are taken from the United Nations database and are measured in billion constant 2010 US dollars. Total investment includes gross domestic fixed capital formation and direct foreign investment. The source of data on net inflows of foreign direct investment is UNCTAD, and it is expressed as a percentage of GDP.

The decomposition model

Kaya (1990) proposed an identity which states that the total emission level of carbon dioxide can be expressed as the product of four factors: emission intensity (emissions per unit of primary energy consumed), energy intensity (primary energy consumption per unit of GDP), GDP per capita and population, that is,

$$CO_2 = \left(\frac{CO_2}{PE}\right) \times \left(\frac{PE}{GDP}\right) \times \left(\frac{GDP}{P}\right) \times P \tag{3.1}$$

Now, the two factors in the previous equation can be extended as

$$\left(\frac{CO_2}{PE}\right) = \left(\frac{CO_2}{FF}\right) \times \left(\frac{FF}{PE}\right) \text{ and} \tag{3.2}$$

$$\left(\frac{GDP}{P}\right) = \left(\frac{GDP}{I}\right) \times \left(\frac{I}{FDI}\right) \times \left(\frac{FDI}{P}\right) \tag{3.3}$$

Using Equations (1), (2) and (3), CO_2 emissions can be decomposed into seven factors.

Hence CO_2 emission in country *i* can be expressed as

$$CO_{2i} = \left(\frac{CO_{2i}}{FF_i}\right) \times \left(\frac{FF_i}{PE_i}\right) \times \left(\frac{PE_i}{GDP_i}\right) \times \left(\frac{GDP_i}{I_i}\right) \times \left(\frac{I_i}{FDI_i}\right) \times \left(\frac{FDI_i}{P_i}\right) \times P_i \quad (3.4)$$

$$= C_i D_i E_i Y_i K_i F_i P_i \quad (3.5)$$

In Equation (3.5), $C_i = \left(\frac{CO_{2i}}{FF_i}\right)$ is the emission intensity of fossil fuel, representing the carbon dioxide emissions per unit of fossil fuel use; $D_i = \left(\frac{FF_i}{PE_i}\right)$ represents the dirty energy intensity, which shows the proportion of fossil fuel use in total energy consumption and captures the energy substitution effect; $E_i = \left(\frac{PE_i}{GDP_i}\right)$ is the energy intensity of the GDP and indicates the energy efficiency parameter; $Y_i = \left(\frac{GDP_i}{I_i}\right)$ is the income intensity of the investment and captures GDP per unit of investment; $K_i = \left(\frac{I_i}{FDI_i}\right)$ is the FDI intensity, indicating the ratio of total investment to FDI; $F_i = \left(\frac{FDI_i}{P_i}\right)$ is the per capita FDI, and P_i is the population. The first three factors capture the extent of the efficacy of technology in curbing emissions.

Based on this structure, the changes in the emissions of a country (DCO_{2i}) from a base year 0 to an end year T can be decomposed into seven different effects, as mentioned previously: change in emissions per unit of fossil fuel (C_{eff}: emission intensity of fossil fuel effect), change in fossil fuel use in total energy use (D_{eff}: dirty energy intensity effect), change in energy consumed per unit of GDP (E_{eff}: energy intensity effect), change in GDP per unit of investment (Y_{eff} : income intensity effect), change in total investment per unit of FDI (K_{eff} : investment intensity off effect), change in per capita FDI effect (F_{eff}: FDI intensity effect) and change in population effect (P_{eff}).

Hence, the change in the emissions of country *i*(ΔCO_{2i}) from a base year 0 to an end year T can be expressed as a sum of the effects of the seven contributing factors (Equation 3.6).

$$DCO_{2i} \equiv CO_{2i}(T) - CO_{2i}(0) \equiv C_{eff} + D_{eff} + E_{eff} + Y_{eff} + K_{eff} + F_{eff} + P_{eff} \quad (3.6)$$

Thus, from Equations (3.2) and (3.3), the emission intensity effect of energy is the sum of effects of emission intensity of fossil fuel and dirty energy intensity, while the per capita income effect is the sum of effects of income intensity of investment, FDI intensity and per capita FDI.

In accordance with the LMDI method in additive form postulated by Ang (2005), the seven contributing factors can be derived from Equations (3.7)–(3.13). Each of the effects is isolated by measuring changes in CO_2 emissions associated with the change in the corresponding variable while fixing the other variables constant with respective values in the base year.

$$C_{eff} = \sum_i L\left(CO_{2i}(T)CO_{2i}(0)\right)\ln(C_i(T)/C_i(0)) \tag{3.7}$$

$$D_{eff} = \sum_i L\left(CO_{2i}(T)CO_{2i}(0)\right)\ln(D_i(T)/D_i(0)) \tag{3.8}$$

$$E_{eff} = \sum_i L\left(CO_{2i}(T)CO_{2i}(0)\right)\ln(E_i(T)/E_i(0)) \tag{3.9}$$

$$Y_{eff} = \sum_i L\left(CO_{2i}(T)CO_{2i}(0)\right)\ln(Y_i(T)/Y_i(0)) \tag{3.10}$$

$$K_{eff} = \sum_i L\left(CO_{2i}(T)CO_{2i}(0)\right)\ln(K_i(T)/K_i(0)) \tag{3.11}$$

$$F_{eff} = \sum_i L\left(CO_{2i}(T)CO_{2i}(0)\right)\ln(F_i(T)/F_i(0)) \quad 3.12)$$

$$P_{eff} = \sum_i L\left(CO_{2i}(T)CO_{2i}(0)\right)\ln(P_i(T)/P_i(0)) \tag{3.13}$$

where logarithmic mean function

$$L(CO_{2i}(T)CO_{2i}(0) = \frac{CO_2(T) - CO_2(0)}{\ln(CO_2(T)) - \ln(CO_2(0))}$$

Results and discussion

Preliminary analysis

The total CO_2 emission of the six countries, Afghanistan, Bangladesh, India, Nepal, Pakistan and Sri Lanka, increased more than 14 times, from 184.03 million metric tons in 1980 to 2674.89 million metric tons in 2015 (Table 3.1). In both years, India accounted for the highest share in emissions (76.3% in 1980 and 89.4% in 2015), followed by Pakistan (16.1% in 1980 and 6% in 2015), Bangladesh (3.8% in 1980 and 2.7% in 2015), Sri Lanka (2.6% in 1980 and 0.7% in 2015), Afghanistan (0.7% in 1980 and 0.8% in 2015) and Nepal (0.2% in 1980 and 0.1% in 2015). During 1980 and 2015, India's emissions increased 17 times, while those from Afghanistan,

Table 3.1 Country-wise values of the factors in 1980 and 2015

Year	*Country*	CO_2 *(MMT)*	*FF (TBtu)*	*PE (TBtu)*	*GDP (Billion$)*	*Total inv. (Billion$)*	*FDI (% of GDP)*	*Population (millions)*
1980	Afghanistan	1.36	19.98	27.35	10.39	1.98	0.25	13.25
	Bangladesh	7.07	115.71	122.91	28.32	5.01	0.05	81.47
	India	140.53	3453.68	3970.85	262.77	53.01	0.04	698.82
	Nepal	0.47	6.03	8.01	4.25	0.70	0.02	14.90
	Pakistan	29.72	534.98	624.60	43.12	9.55	0.27	78.07
	Sri Lanka	4.88	73.00	88.21	13.55	3.99	1.07	15.04
	Total	184.03	4203.38	4841.93	362.40	74.25	0.11	901.54
2015	Afghanistan	22.82	106.90	130.22	20.89	2.29	0.88	33.74
	Bangladesh	72.95	1265.34	1272.36	155.58	45.14	1.45	161.20
	India	2391.37	28709.61	30985.71	2293.06	748.00	2.09	1310.96
	Nepal	4.40	55.54	94.44	20.12	5.26	0.24	28.66
	Pakistan	162.28	2564.44	2942.31	212.37	28.11	0.60	189.38
	Sri Lanka	21.07	292.00	350.98	76.36	20.25	0.84	20.71
	Total	2674.89	32993.82	35776.02	2778.38	849.05	1.89	1744.64

Source: Based on authors' calculations

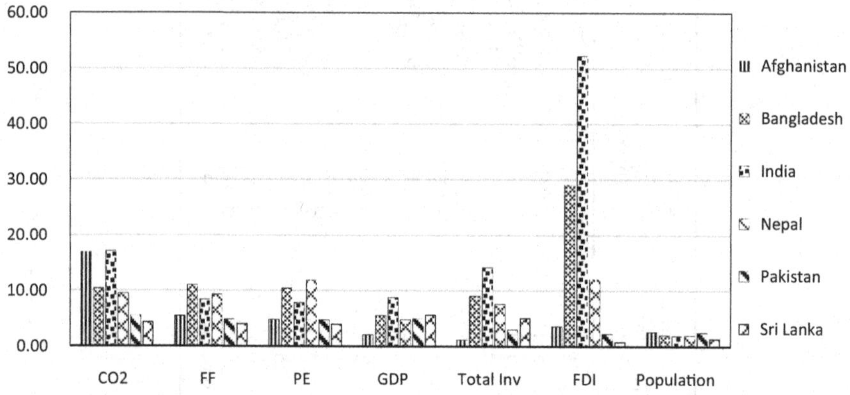

Figure 3.1 Growth of the factors (expressed as number of times that of the base year)

Source: Based on authors' calculations

Bangladesh, Nepal, Pakistan and Sri Lanka grew16.7, 10.3, 9.3, 5.4 and 4.3 times, respectively (Figure 3.1).

Apart from emissions, India, Pakistan and Bangladesh had the highest values for all other variables in absolute terms as well; however, the scenario was different in terms of growth rates. The total consumption of primary energy of the countries rose by about seven times, from 4841.93 trillion Btu in 1980 to 35,776.02 trillion Btu in 2015. The maximum growth in primary energy use was experienced by Nepal (11.7 times), followed by Bangladesh (10.3 times), India (7.8 times), Afghanistan (4.8 times), Pakistan (4.7 times) and Sri Lanka (3.9 times). The use of fossil fuel in the six countries increased nearly eight-fold, from 4203.38 trillion Btu in 1980 to 32,993.82 trillion Btu in 2015. The surge in fossil fuel consumption was the highest in Bangladesh (10.9 times), followed by Nepal (9.2 times), India (8.3 times), Afghanistan (5.3 times), Pakistan (4.7 times) and Sri Lanka (4 times). Not only did fossil fuel consumption increase, per capita fossil fuel consumption also shot up in the countries, the highest being in Bangladesh (5.5 times), followed by Nepal (4.8 times), India (4.4 times), Sri Lanka (2.8 times), Afghanistan (2.1 times) and Pakistan (2 times).

The total GDP of the countries increased 7.6 times during the period, with the highest improvement being experienced by India, and then Sri Lanka, Bangladesh, Pakistan, Nepal and Afghanistan. Investment increased by 11.4 times in aggregate; India experienced a spectacular rise in investment by 14 times, while that in other countries was far behind, with the corresponding figures for Bangladesh, Nepal, Sri Lanka, Pakistan and

Afghanistan being 9, 7.5, 5, 2.9 and 1.1 times, respectively. During this period, there was a remarkable rise in total foreign direct investment by 17 times, India achieving an enormous rise in FDI by 52 times, followed by Bangladesh (29 times). In Nepal, Afghanistan, Pakistan and Sri Lanka, the inflow of foreign capital increased 12, 3.52, 2.2 and 0.7 times, respectively. The population nearly doubled in the six countries, the maximum growth being experienced by Afghanistan, followed by Pakistan, Bangladesh, Nepal, India and Sri Lanka.

Figure 3.1 shows that in Afghanistan, emissions growth has been extensive despite very low growth in GDP. Sri Lanka experienced the lowest fossil fuel consumption and emissions, while there has been impressive growth in GDP. The rate of growth of emission was much less than the high growth of primary energy use in Nepal. To explain these apparent contradictions, a decomposition analysis is undertaken.

Decomposition analysis

At the aggregate level, the increase in emissions was due to contributions of the rise in per capita income, emission intensity of energy and population, each accounting for 53%, 25% and 24% of emissions growth, respectively; however, the energy intensity effect was able to offset only 2% of the emission-enhancing effects (Table 3.2). The country-wise break-up shows that per capita income had the most significant role in raising emissions, accounting for 43%, 54%, 40% and 41% in Bangladesh, India, Nepal, and Pakistan; in Sri Lanka, it accounted for 96% of emissions. However, in Afghanistan, it offset about 8% of the emission-raising effects of the other factors. Population had an emission-increasing effect in all the countries, accounting for 33%, 29%, 22%, 29% and 21% of the rise in emissions in Afghanistan, Bangladesh, India, Nepal and Sri Lanka, respectively; for Pakistan, the corresponding figure was remarkably high, at 52%.

The decomposition of the per capita income effect indicates that income intensity of investment and investment intensity of FDI had offsetting effects of about 8% and 63% of the remarkably high positive effect of FDI per capita. While all individual countries had high emission-raising effects of FDI per capita, income intensity of investment had negative effects in Bangladesh, India and Nepal; negative effects were observed regarding investment intensity of FDI in all the countries, except Sri Lanka.

The performance of the countries with respect to emission intensity of energy and energy intensity of GDP has been bleak. While Bangladesh and Nepal had negative emission intensity effects that neutralized 0.1% and 9.9% of the rise in emissions, emission intensity had a predominantly significant positive effect in most of the countries, implying lack of technological improvements. India and Pakistan experienced offsetting effects of energy intensity of GDP to the extent of only 3.9% and 2.6% of the

Table 3.2 Decomposition results

Country	*Emission effect of energy*			E_{eff}	*Per capita income effect*				P_{eff}	ΔCO_2
	C_{eff}	D_{eff}	*Total*		Y_{eff}	K_{eff}	F_{eff}	*Total*		
Afghanistan	8.70	0.89	9.59	6.56	4.22	-13.89	7.87	-1.8	7.11	21.46
Bangladesh	-1.63	1.55	-0.08	17.88	-13.94	-82.91	125.68	28.83	19.26	65.88
India	568.95	50.23	619.18	-88.79	-381.65	-2702.85	4305.32	1220.82	499.63	2250.84
Nepal	0.04	-0.43	-0.39	1.60	-0.82	-4.01	6.41	1.58	1.15	3.93
Pakistan	10.16	1.36	11.52	-3.48	40.24	-102.89	117.96	55.31	69.21	132.55
Sri Lanka	0.85	0.06	0.91	-3.86	1.17	1.45	12.98	15.6	3.55	16.19
Total	587.07	53.65	640.72	-70.08	-350.78	-2905.10	4576.22	1320.34	599.90	2490.86

Source: Based on authors' calculations (in terms of million metric tons of CO_2)

emission-raising effects of the driving forces. However, Sri Lanka succeeded in offsetting 23% of emissions through improvements in energy use.

At the aggregate level, decomposition of the emission intensity of energy shows that although both emission intensity of fossil fuel and share of dirty energy had positive effects, the former was the most contributing factor (about 92%). In Bangladesh, emission intensity of fossil fuel had a strong negative effect that offset the positive dirty energy intensity effect. All other countries experienced a high positive effect of emission intensity of fossil fuel. Except for Nepal, the dirty energy intensity effect was also positive for the countries.

Although previous studies[4] show that emission intensity and energy intensity are catalytic in curbing rising emissions due to income and population effects, this study points towards their insignificant role. This entails further examination of the nature of change in energy mix and the emissions per unit of fossil fuel to understand the deviation.

Study of the energy mix and emission intensity of fossil fuels

A remarkable fact is that although Nepal had the highest growth rate in primary energy use as well as a considerable increase in fossil fuel, its emission growth rate has been much less. On the other hand, Afghanistan experienced the second-highest growth rate in emissions, despite the rate of increases in primary energy use and fossil fuel being much lower. In order to gain a clear insight into the linkage between emissions and primary energy use, we consider the composition of primary energy in terms of fossil fuels (coal, petroleum and natural gas) and clean renewable fuel.

Table 3.3 and Figure 3.2 show that within their respective fuel mix, Afghanistan has inclined towards a higher rate of use of coal and a lower rate of use of petroleum, followed by natural gas; Bangladesh has emphasized the rate of use of natural gas more, reducing that of petroleum; Nepal remarkably

Table 3.3 Composition of primary energy (in percentage) in 1980 and 2015

Country	*Coal*		*Petroleum and other liquids*		*Natural gas*		*Nuclear, renewable and other*	
	1980	*2015*	*1980*	*2015*	*1980*	*2015*	*1980*	*2015*
Afghanistan	9.00	22.58	54.48	54.32	7.61	4.29	28.91	18.80
Bangladesh	0.40	3.10	55.35	18.18	38.30	78.12	5.95	0.60
India	49.69	55.54	35.43	29.63	1.52	6.56	13.36	8.27
Nepal	17.53	4.93	52.53	54.02	0.00	0.00	29.93	41.05
Pakistan	5.79	6.66	35.12	37.53	44.72	42.75	14.37	13.06
Sri Lanka	0.00	15.59	82.71	67.10	0.00	0.00	17.29	17.30

Source: Based on authors' calculations

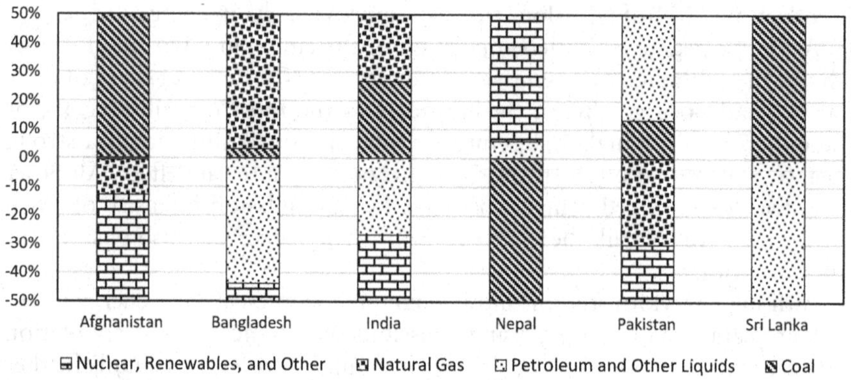

Figure 3.2 Percentage change in composition of primary energy from 1980 to 2015

Source: Based on authors' calculations

emphasized clean energy by lowering its use rate of coal; India reduced its rate of use of petroleum and clean energy while raising the use of coal and natural gas; Pakistan enhanced the rate of use of coal and petroleum while lowering the use rate of natural gas and clean energy; Sri Lanka emphasized a higher rate of use of coal, reducing the use rate of petroleum. It is evident from these changes in energy composition that the stated higher growth rate of emissions in Afghanistan is the result of a drastic rise in the rate of use of coal, while the impressive shift towards renewable clean fuel in Nepal brought down the emission growth rate. However, not only a shift from fossil fuels to renewable clean energy and vice versa but shifts within fossil fuels also significantly drive the magnitude of emissions, as in the case of the other four countries.

Table 3.4 shows that coal is about 36% and 78% more polluting than petroleum and natural gas, respectively, while petroleum is about 32% more polluting than natural gas. This perhaps explains why countries like Afghanistan, India, Pakistan and Sri Lanka, which accentuated their rate of use of coal, experienced higher emission intensity of primary energy. On the other hand, Nepal and Bangladesh made a shift in fuel mix towards clean energy and the least polluting natural gas, respectively, and consequently succeeded in bringing down emission intensity.

From the results of our analysis, the following facts are evident:

- In almost all the South Asian countries, rise in per capita income was the predominant reason for escalating emissions. Higher income not only led to a rise in total energy consumption, it was accompanied by more energy-intensive consumption as well. However, the technological factors had insignificant effects due to the high emission intensity of energy and lack of judicious energy use.

Table 3.4 Emission intensities

Country	*Emission intensity of primary energy (mg/Btu)*		*Emission intensity of fossil fuel (mg/Btu)*		
	1980	*2015*	*Coal*	*Petroleum and other liquids*	*Natural gas*
Afghanistan	48.16	61.40	93.30	70.05	53.06
Bangladesh	57.87	57.77	94.71	73.67	53.06
India	73.38	75.51	96.83	61.41	53.88
Nepal	53.73	42.27	97.20	69.38	-
Pakistan	52.72	56.69	95.07	73.58	53.18
Sri Lanka	55.27	63.25	93.43	72.52	-

Source: Based on authors' calculations

- Although higher total investment had a major offsetting effect (except Sri Lanka), rising FDI had a significant role in rising emissions in all countries.
- India had been the topmost emitter in the region, both in absolute terms as well as with respect to growth rate. Although energy intensity has produced some favourable effects, higher emission intensity is a concern for India.
- Population increase had a predominant role in raising emissions in Afghanistan, while the per capita income effect had offsetting effects.
- In Bangladesh, energy intensity coupled with population growth were significant factors of emissions. With higher energy use per person, an increase in population has an amplified effect on emissions. Both income and a shift in fuel mix towards natural gas had a favourable effect.
- Nepal had the lowest share in CO_2 emissions in the region; a change in energy source towards clean fuel and the income effect contributed to offsetting emissions.
- Population and income intensity were major players in surging emissions in Pakistan.
- Sri Lanka experienced the lowest growth in emissions during the period. Per capita income was the most important driving force behind rising emissions, and prudent energy use had significant offsetting effects.

With invigorated global concern regarding greenhouse gas emissions resulting in the perils of climate change, a series of deliberations are being made, often resulting in treaties at the international level. In adherence to the agreements in the climate change negotiations, a majority of the countries have adopted environmental policies to reduce emissions.

Policies relating to emissions cuts

After identification of major emitting countries in South Asia and the driving forces of emission, it becomes imperative to ascertain the mechanisms for arresting rising emissions. While higher income cannot be compromised for environmental concerns, population control is a long-term phenomenon with slower effects on emissions, and restriction of FDI in the globalized world may severely affect the growth of these underdeveloped host countries. The onus of emissions reductions then rests on technological factors through reductions in emissions and energy intensities. These can best be accomplished by framing rigid environmental regulation and properly implementing them. It would be worthwhile to review the nature and extent of environmental policies adopted by the countries. The review is based on the UNEP Report, 2015.

Afghanistan: The predominant source of CO_2 emission is transport, attributable to rampant use of old and ill-maintained vehicles, poor quality fuel use and illegal import of used vehicles. The share of industry in the GDP is only 6%; nonetheless, the chemical and metal industries and power plants use poor-quality fuel and contribute as the primary emitting industries. The legal framework for environmental regulation is extremely weak. There exists an air quality monitoring station, guided by the national ambient air quality standards framed per WHO guidelines, and a national air quality policy steered by the Clean Air Regulation, 2010. However, there is lack of monitoring, leading to inadequate implementation of the policies. There is no emissions regulation for industries, no limit on vehicle emissions and a lack of objective policies for promotion of renewable energy, energy efficiency initiatives and emission prevention technologies.

Bangladesh: Brick kilns, transport, cement and metal smelting are the major sectors that contribute to emissions. There are 11 continuous air monitoring stations that identify the extent of pollution in the cities and develop an air quality index for public awareness. There is no specific air quality legislation – the issue is addressed in the Bangladesh Environment Conservation Act, 1995, and Environment Conservation Rules, 1997. The monitoring of emissions in industries is inadequate since the Department of Environment lacks manpower and technical knowledge for effective emissions monitoring. Lack of information on emissions affects the enforcement of legislation. Small installations are regulated in the form of requirements for environmental clearance, whereby industries are categorized into four groups in accordance with their environmental performance. Although policies exist for promotion of investment in renewable energy, financial and regulatory support are inadequate. The Bangladesh Bank provides soft loans to industries for using low-emission technology; the Department of Environment conducts awareness programs to motivate industries to switch towards less polluting technology; there are limits to vehicular emissions, restrictions on import of used cars and actions have been undertaken to promote non-motorized transport. However, the regulations lack stringent implementation.

India: Emissions are covered under the Air (Prevention and Control of Pollution) Act, 1981. The National Ambient Air Quality standard is followed and monitored by an air quality monitoring system. Petroleum refineries and the chemical, textile, steel, cement and basic metal industries are the major industrial polluters. A number of policies exist for encouraging investment in renewable energy, like exemptions from customs and excise duties and state-level tax exemptions. The Indian Renewable Energy Development Agency promotes and finances renewable energy projects, and there are national and state programmes to promote solar projects. There are four regional cleaner production centres to create awareness, provide training and so on and a credit-linked capital subsidy scheme for technology upgrades of small-scale industries. There are regulatory mechanisms for industrial emissions, but the actions to ensure compliance remain questionable. The vehicle emissions limit follows the Euro 3 mandate and Euro 4 for 11 cities, and actions are being taken to promote non-motorized transport.

Nepal: Although Nepal improved its emissions level over our study period, it ranked 177th out of 178 countries for environmental quality in the 2014 Environmental Quality Index. Metal manufacture and brick kilns are the most polluting industries; old and poorly maintained vehicles with fuel adulteration add to the emissions. Nepal does not have a national air quality policy, with no specific law to address air pollution. However, it maintains standards limiting emissions from industries; more commendably, there are significant policies relating to energy use. The Renewable Energy Subsidy Policy, 2000, provides grants for solar, hydro and biogas investments. The Industrial Energy Management Project that started in 1994 has been conducting energy audits of industrial boilers and equipment – it has been found that about two-third of the industries audited have been investing in energy efficiency equipment. The Industrial Enterprises Act grants a 50% reduction in taxable income for industries that invest in equipment or processes that control pollution in three years. However, the monitoring and enforcement to ensure compliance with the regulations are considerably weak. The vehicle emissions limit follows the Euro 3 mandate, although pre-euro vehicles are also in use.

Pakistan: It is the most urbanized country in South Asia, with pollution mainly due to vehicular and industrial emissions. Cement, fertilizer, sugar and steel are the industries that have the highest impact on emissions. There was an air quality monitoring system, which has become non-operational due to lack of maintenance. There is no national air quality policy – emissions are under the ambit of the Pakistan Environmental Protection Act that provides prohibitions, penalties and an enforcement mechanism. The Pakistan Clean Air Program provides a list of interventions with regard to emissions. There is no vehicular emissions limit. Nonetheless, there are some policies that address efficient use of energy, like waiver of import duties and other tax benefits on purchase of equipment that promote renewable energy use and promotion of renewable energy by the Pakistan Council of Renewable Energy Technologies through training, research, policies and so on. The Energy Efficiency and Conservation Bill, 2014, has been catalytic in taking initiatives for setting up energy efficiency standards and labelling for some products. The Pollution Charge Rules 2001 laid out the formulas for pollution taxes but were hardly implemented due to resistance from powerful lobbies.

Sri Lanka: Emissions in urban areas are mainly due to vehicles and thermal power plants. The cement, textile, petroleum refinery and agro-processing industries have the potential to have significant impact on emissions. Although there is no national air quality policy, there are three monitoring stations and legislation like the Clean Air Action Plan 2000. The region lacks personnel and capacity to tackle pollution outside Colombo, with the Central Environmental Authority having limited regulatory and enforcement powers. Industrial pollution is regulated through emission licensing and the National Industrial Pollution Management Policy Statement. There are provisions for subsidies and tax reductions for purchase of equipment promoting energy efficiency; the Cleaner Production Policy and National Cleaner Production Centre provide incentives for clean production; however, economic or incentive-based instruments for encouraging pollution control technology are absent. The vehicle emissions limit follows the Euro 1 mandate, and vehicles older than four years are banned.

The Environmental Performance Indicator, 2018, prepared by Yale University indicates that Sri Lanka is the best performer among the South Asian countries, ranking sixth among all the Asian countries, with the rankings of Afghanistan, Pakistan, Nepal, India and Bangladesh being 22, 23, 24, 25 and 26, respectively. However, the low emissions in Sri Lanka can be predominantly attributed to the sluggish rise of its GDP. Similarly, falling income has had an emissions-reducing effect in Afghanistan. The lower-performing countries have serious pitfalls in regulatory policies and/or their enforcement so that the effects of emissions-increasing driving forces could not be sufficiently compensated for.

Conclusion

The South Asian countries have experienced massive growth in CO_2 emissions in the last few decades. A plethora of economic and demographic changes have occurred in the region as well that might have conspicuous implications on the emissions level. This chapter examined the contributions of seven determining factors in the change in emissions level from 1980 to 2015 in six countries of South Asia using log mean divisia index method. The results indicate that a rise in per capita income was a major source of emissions in the region. This is because higher income prompts higher energy consumption. Although the per capita energy consumption in South Asia is still low compared to other advanced countries, a steep rise in its value has been observed in the recent period. Since energy consumption is mainly sourced from fossil fuel in the region, it adds to emissions. A further decomposition reveals that inflow of FDI per capita had the dominant emission-enhancing effect. However, an interesting point revealed from the findings is that higher total investment has been instrumental in lowering emissions. This indicates that these countries, eager for rapid economic growth, allow less efficient technology and dirty industries from foreign investors to grow within their territories. No significant gainful effect of technology pertaining to emissions and energy intensity could be observed.

The results point towards a serious lacuna in government interventions in order to arrest escalating emissions. Since technological factors seem to have an insignificant contribution in reducing emissions, efforts for reduction in energy use and energy substitution of fossil fuels with cleaner energy sources need to be geared up. A radical shift towards renewable energy may not be feasible due to two reasons. First, the initial costs of renewable energy are quite high; in this case, gradual substitution of the highest-polluting fuel by less-polluting ones can be a viable alternative. Second, conventional energy employs a number of people in South Asia; energy substitution may initially pose some displacement problems, but the displaced workers may be trained and shifted to other sectors. There are also opportunities for energy trade within South Asia, where the renewable energy resource surplus countries can benefit from export-led growth, which would simultaneously meet demands in energy deficient counties. For example, hydropower potential in Nepal, wind power potential in Afghanistan and solar power potential in India can be exploited, traded and used for any additional energy demand.

Stringent environmental regulations are likely to deter the migration of foreign-owned "dirty" industries. Moreover, foreign firms can be induced to acquire and transfer environmentally sound technology not only by formulating strong environmental regulations but by providing them various emissions-related financial incentives. However, mere formulation of policies to curb emissions can hardly be successful without close monitoring and stringent enforcement mechanisms, which seem to be inadequate in these countries. All these efforts can be expected to ensure the economic growth and inflow of FDI in the developing countries of South Asia without compromising their environmental effects.

Notes

1 The annual population growth rate of 1.24% in South Asia is higher than the corresponding world figure of 1.15% in 2017; nonetheless, it has declined from 2.40% in 1980 (according to World Bank data).

2 Copeland and Taylor (1994).

3 The other two countries of South Asia, Bhutan and Maldives, could not be included in the study on account of unavailability of adequate matching data.
4 Lee and Oh (2006); World Bank (2007); Pani and Mukhopadhyay (2010).

References

Ang, B. W. 'The LMDI Approach to Decomposition Analysis: A Practical Guide.' *Energy Policy,* 33(2005): 867–871.

Cansino, J.M., Sánchez-Braza, A., and Rodríguez-Arévalo, M. L. 'Driving Forces of Spain's CO_2 Emissions: A LMDI Decomposition Approach.' *Renewable and Sustainable Energy Reviews,* 48 (2015): 749–759.

Casey, G., and Galor, O. 'Population Growth and Carbon Emissions.' Working Paper No. 22885, National Bureau of Economic Research, Cambridge, 2016.

Coondoo, D and Dinda, S. 'Causality between Income and Emission: A Country Group Specific Econometric Analysis.' *Ecological Economics,* 40(3) (2002): 351–367.

Copeland, B.R., and Taylor, M. S. 'North-South Trade and the Environment.' *Quarterly Journal of Economics,* 109(3) (1994): 755–787.

Crowley, T.J. 'Study Faults Humans for Large Share of Global Warming.' *The New York Times,* July 14, 2000.

Engleman, R. *Profiles in Carbon: An Update on Population, Consumption and Carbon Dioxide Emission.* Washington DC: Population Action International, 1998.

Grossman, G. M., and Krueger, A. B. 'Environmental impacts of a North American Free Trade Agreement.' In P. Garber (Ed.), *The U.S.–Mexico Free Trade Agreement.* Cambridge, MA: MIT Press, 1994.

Hang, G., and Yuan-sheng, J. 'The Relationship between CO_2 Emissions, Economic Scale, Technology, Income and Population in China.' *Procedia Environmental Sciences,* 11(2011): 1183–1188.

Harte, J. 'Human Population as a Dynamic Factor in Environmental Degradation.' *Population and Environment,* 28(4–5) (2007): 223–236.

Holtz-Eakin, D., and Selden, T.M. 'Stoking the Fires? CO_2 Emissions and Economic Growth.' *Journal of Public Economics,* 57 (1995): 85–101.

Jiang, L., and Hardee, K. 'How Do Recent Population Trends Matter to Climate Change?' *Population Research Policy Review,* 30(2) (2011): 287–312.

Kaya, Y. 'Impact of Carbon Dioxide Emissions Control on GNP Growth: Interpretation of Proposed Scenarios.' Paper presented to IPCC Energy and Industry Sub-Groups, Response Strategies Working Group, 1990.

Lee, K., and Oh, W. 'Analysis of CO_2 Emissions in APEC Countries: A Time-Series and a Cross-Sectional Decomposition Using the Log Mean Divisia Method.' *Energy Policy,* 34 (2006): 2779–2287.

Lipford, J.W., and Yandle, B. 'Environmental Kuznets Curve, Carbon Emissions and Public Choice.' *Environment and Development Economics,* 15(4) (2010): 438–471.

Mousavi, B., Lopez, N.S.A., Biona, J. B. M., Chiu, A.S.F., and Blesl, M. 'Driving Forces of Iran's CO_2 Emissions from Energy Consumption: An LMDI Decomposition Approach.' *Applied Energy,* 206 (2017): 804–814.

Pani, R., and Mukhopadhyay, U. 'Identifying the Major Players behind Increasing Global Carbon Dioxide Emissions: A Decomposition Analysis.' *The Environmentalist,* 30(2010): 183–205.

Shafik, N. 'Economic Development and Environmental Quality: An Econometric Analysis.' *Oxford Economic Papers,* 46 (1994): 757–773.

Shafik, N., and Bandyopadhyay, S. 'Economic Growth and Environmental Quality: Time Series and Cross-Country Evidence.' Background Paper for World Development Report. The World Bank, Washington, DC, 1992.

Shi, A. 'Population Growth and Global Carbon Dioxide Emission.' Development Research Group, The World Bank Paper presented at IUSSP Conference in Brazil/ Session–s09, 2001.

Wang, W.W., Zhang, M., and Zhou, M. 'Using LMDI Method to Analyze Transport Sector CO_2 Emissions in China.' *Energy,* 36(10) (2011): 5909–5915.

World Bank. *Growth and CO2 Emissions: How Do Different Countries Fare?* Washington, DC: The Environment Department, World Bank, 2007.

Yeo, Y., Shim, D., Lee, J.D., and Altmann, J. 'Driving Forces of CO_2 Emissions in Emerging Countries: LMDI Decomposition Analysis on China and India's Residential Sector.' *Sustainability,* 7(2015): 16108–16129.

Yii, K.J., and Geetha, C. 'The Nexus between Technology Innovation and CO_2 Emissions in Malaysia: Evidence from Granger Causality Test.' *Energy,* 105(2017): 3118–3124.

York, R. 'Demographic Trends and Energy Consumption in European Union Nations, 1960–2025.' *Social Science Research,* 36(3) (2007): 855–872.

Part II

Climate change–induced human migration

Concerns for environmental refugees

4 Nature and characteristics of climate change–induced human migration in South Asia

Manisha Deb Sarkar

Human migration commonly refers to the mass exodus of people from their place of habitat in search of a new homeland. The world has witnessed migration of different groups of people since ancient times. Historical evidence tells us that migration was a very common phenomenon of people irrespective of their spatial or cultural identity. Archaeological findings at places are good evidence of mass movement of people in the geological past. It is also a historical and anthropological truth that the present worldwide distribution of people migrated out from specific regions in Africa to elsewhere (Korisettar et al., 2015).

Change of habitat often poses a question on the change of the quality of liveability. People migrate either individually or en masse due to many reasons. From a demographic perspective, the human migration process involves a number of elements based on the 'origin' and 'destination'. Leaving apart others, climate-induced mass migration, internal or across the border of countries or continents, permanently or semi-permanently, is the theme under consideration here. This type of mass migration of people depends on large-scale long-term changes of liveability and their dependence on the immediate surroundings. Changes in the liveable surroundings depend largely on a major deterministic factor like climate. Climate defines how to live, grow and develop. People become habituated to the climate they live with or even adapt to its slow changes. But when changes are abrupt and severe, it becomes devastating. If this suddenness happens once in a while, people try to cope with it. But if it continues for a long time, they think of leaving. Presently, climate change affecting the liveability of people has become the gravest challenge at all spatial levels. The pace of change of climate can be classified into two kinds – a) the fast or sudden type and b) the slow type. Basically, the first one is a climatic event which involves flood, cyclone, river erosion and so on, and the second one is a climatic process that involves coastal erosion, sea-level rise, salt water intrusion, rising temperature, changing rainfall pattern, drought and the like (Rabbani et al., 2016).

In this context, this discussion endeavours to find the conditional similarities or dissimilarities that might have happened leading to a mass exodus of people from their centuries-old habitation. As happened in past ages due to climate change, with the present condition of changing climate, it is a question of how the victimised people react, from the natural to anthropogenic factors. The socio-political and socio-economic scenario has completely changed. It is, therefore, a matter of management how to perceive and analyse the behavioural changes of the affected people and their power of adaptability and resilience and the direct and indirect forces highlighting the 'push' or 'pull' factors in order to contemplate their lives and livelihood by formulating and implementing emphatic policies in favour of them at a global level (Piguet et al., 2011).

Climate-induced human migration is not new

Past episodes suggest that climate change from a global perspective is not new. In South Asia, it has vastly affected people living here since ancient times. The major landmass of South Asia is geographically located within tropical and sub-tropical thermal zones (a region with high variability of temperature and humidity), where the lofty Himalayas in the north and the oceanic expanse in the south play a steering role (Kathayat et al., 2017). It is a zone of constant atmospheric turmoil in response to the variations that control it. In short, climate change is not just fluctuations of the atmospheric conditions; rather, it is an interactive result of land, water bodies and the atmospheric gaseous mass. On a geological time scale, during the historical past, there are indicators of climate change over time (Khan and Gangawala, 2011).

Migration constitutes a huge part of human history. Paleo-climatologists agree that early humans started out from Africa and began to migrate out of the continent to different parts of Europe and Asia around 100,000 years ago. According to geoscientists at the University of Arizona, humans migrated out of Africa as the climate shifted from wet to dry about 60,000 years ago. Tierney and others found that around 70,000 years ago, the climate in the Horn of Africa shifted from a wet phase called *Green Sahara* to even drier than the region is now. The region also became colder (Tierney et al., 2017). As the environment was deteriorating from wet to dry, that forced people to migrate. Climate-induced migration from Africa was the crucial reason for the present global distribution of the population (Science Daily, 2017). In fact, similar situations had occurred many times in the global climatic scenario through the ages. Desertification or extreme aridity in this thermal zone had a big hand in the collapse of many ancient civilisations. Egypt, the early Bronze Age civilisations of Greece and Crete and the Akkadian empire of Mesopotamia were all victims of extended droughts (Shaowu, 2010; Sarkar et al., 2016).

In South Asia, the most significant example is the Indus Valley civilisation in Harappa and Mohenjo-Daro. In a recent publication in *Nature*, researchers from the Indian Institute of Technology (IIT) at Kharagpur, India, found profound evidence of climate change and human migration in India. Previously, foreign invasions, social instabilities and declines in trade had been given as reasons for the decline of this ancient civilisation. But these scientists showed that it was climate change (consistent drought for 200 years) that played the decisive role in large-scale migration (Moudgil, 2017). It is the conflict between people and nature that led to migration in search of a better land and climate. Researchers over the years have found that the Indian monsoon was weak until 9000 years ago, and the people depending on rainwater were only pastoralists or small-scale farmers (Moudgil, 2017). Later, with the intensification of rainfall and proliferation of river water, intensive agriculture was possible. There are signs of many ancient habitations being excavated in present-day Haryana through Rajasthan (Misra, 1984). From a study on stalactite/stalagmite formations in Sahiya cave, located about 200 km north of Delhi in Uttarakhand, high-resolution oxygen isotope records of over the last 2000 years revealed that it is strongly influenced by spatially integrated upstream changes in the Indian summer monsoon (ISM) circulation dynamics and rainfall amounts (Kathayat et al., 2017). Researchers have envisaged a relationship between ISM variability and civilisation changes in this region.

Human adaptation to climate change in historical times

The dependence of the Harappan people on climate is quite explicable. These people were agriculturists, and as the climate turned drier, the paucity of water threatened their food supply. The monsoon started declining around 7000 years ago, and a harsh drought 4000 years ago left even the large rivers dry (Moudgil, 2017).

A study shows a

> Dramatic decrease in the availability and the seed density of wheat and barley in the later Harappan period (3,850–3,250 years ago) as compared to the early and mature Harappan periods (5,250–3,850 years ago).
>
> (Moudgil, 2017)

Researchers from the University of Cambridge in the United Kingdom and Banaras Hindu University in Uttar Pradesh worked together in northwest India between 2007 and 2014. They studied the dynamics of adaptation and resilience in the face of a diverse and varied environmental context, using the case study of the Indus Civilisation (3000–1300 BC) (Cameron et al., 2017). The changed climate was somehow being adapted to by the Indus Valley civilisation by changing the variety of food grains. Water-loving crops were replaced with harder varieties, for example, from millets and rice to wheat and barley, which can germinate in winter and spring. Experts say millets, rice and tropical pulses were used in the pre-urban and urban phases of the ancient culture. A shift in cropping was also noticed by archaeologists from a single season–based strategy to one dependent on both seasons (Moudgil, 2017). This shift could have affected the economy and trade patterns of the people, which were quite similar to those of the Mesopotamians. Kumar (2014) mentioned the practices of climate change adaptations and mitigations by the application of indigenous knowledge and skills in India. He took examples from north-western India. As the region is dry and prone to aridity, the basic methodology was effective utilisation of scarce water resources.

The IPCC Report, 2014, on the present situation of climate change, has identified many adaptation options in South Asia. Already this part of the world has experienced many adverse changes in climatic situations and has also tried to adapt at local or regional levels. But situations could worsen, and that might need far more effective adaptations in the future. In the agricultural sector, for example, the cultivation of different crops needs to be adapted differently at different locations in the region. As the IPCC Report suggested, the policies for adapting to climate change could build on the local and indigenous coping strategies of farmers who have been adapting to climatic risks for generations. Breeding crop varieties suited to high temperatures could also be a promising option for adapting to climate change in South Asia (IPCC Report, 2014).

The point to be noted here is that the strategy of adaptation by the farmers to combat climate change looks the same even 5000 years after the existence of the Indus valley civilisation. Scientists have noted that the Indus valley civilisation was primarily agro based. India's economy today is based on agriculture, and around 75% of the people are directly or indirectly dependent on it, which, again, depends on the extreme variability of the monsoon. Though people tried to shift to resilient varieties of crops, a long discontinuation of rainfall caused a food scarcity. This ultimately forced the people to migrate. The length of the time period (a slow process) of the variability of monsoon rainfall and the reluctance of the inhabitants to leave their ancestral habitation are of the utmost importance. It was at an extreme point of time when they were bound to take their final decision. Dryness is a slow process; sometimes it is imperceptible over one generation. But if severity continues, a drastic decision is the obvious result. Today, by observing rainfall variability in some parts of the country, there is a probability of repetition of the same historic event. The water resource situation of the country is also not very promising.

Not such a wonderful world: the 'climatic hot spots' in South Asia

Climate change today is more a human-induced phenomenon than a natural process. Official documents (IPCC, 2014) indicate that there have been warmer days and fewer cold days since 1950, and the frequency of heat waves has increased in South Asia. The scientific community has been on a quest for evidence in establishing the biophysical extent and nature of anthropogenic climate change. The complex interactions between different meteorological and social factors have made climate change–induced human migration a complex prognostic issue. A sharp rise of temperature is indicative of that. The high density of the population living in South Asia often experiences climate severity in the form of cyclones, tornadoes, heat and cold waves, hailstorms and thunderstorms that threaten them with immense vulnerability. There are always potential risks of adaptation failure and low resilience from the victimised parties. These act as 'push factors' for migration with their repetitive and increased frequency in a place. South Asia ranks high on list of the most threatened regions of the world. Researchers noted in the journal *Science Advances* that 4% of the South Asian population will experience dangerous environmental conditions with high temperature and high humidity by 2100 (Worland, 2017). This will primarily affect freshwater resources and agricultural production, creating food scarcity, and the adverse situation will not only jeopardise the economy but will also affect human health and nutrition. Declines in agricultural production and productivity have started bringing declines in food security. The frequency of natural calamities has increased the susceptibility of the region's already-vulnerable population (Kaur and Harpreet, 2018). South Asia is particularly vulnerable to climate change for many reasons. A high density of population lives in the 'climatic hot spots' or in severe climatic zones with unhealthy temperatures and variable precipitation. The ADB Report (2009) has identified 'climatic hot spots' in the

- Low-lying coastal areas (on the Arabian and Bay of Bengal coasts);
- Deltaic regions, which are often the region's most densely populated rural areas;
- Urban centres (coastal megacities like Karachi, Mumbai, Chennai, Kolkata, Dhaka);
- Low-lying small islands (Maldives); and
- Semi-arid or low-humidity regions.

The 'climatic hot spots' have already started experiencing the events of climate change in the form of coastal inundation, desertification and other environmental degradations. Migration has emerged as an important survival and adaptation strategy. Around 15 million people living in the coastal zones of Bangladesh alone are expected to experience forced migration due to this changing situation (Kaur and Harpreet, 2018). However, it is not easy to decipher who the climate-induced migrants are. Some major migration corridors have been identified by Kaur et al. where transboundary migrations have taken place and are still continuing. These are, for example, between Bangladesh and India, Afghanistan and Pakistan, India and Pakistan or Nepal and India. In India alone, there are currently (2018) about

3,230,025 and 810,172 migrants from Bangladesh and Nepal, respectively. Pakistan has about 2,326,275 Afghani migrants. As climate change is largely invisible, this has mostly remained a silent issue. The Intergovernmental Panel on Climate Change Fifth Assessment Report (IPCC AR5) anticipates that these are likely to be felt more severely in future (Anderson et al., 2016).

Generally climate change lasts for several years. Hence, increasing climatic adaptability for the next few decades could cost billions of dollars for stability and survival. Donor countries promised an annual $100 billion by 2020 through the Green Climate Fund for developing countries to adapt to climate change (Kumar, 2014). By 2099, the world is expected to be hotter by an average of between 1.8°C and 4°C than it is now. Large areas are expected to become drier – the proportion of land with constant drought is expected to increase from 2 to 10% by 2050 (Farrant et al., 2006). The proportion of land suffering from extreme drought is predicted to increase from 1% at present to 30% by the end of the 21st century (Carling, 2006).

The World Bank Report (2013) predicted that

- A 2°C rise in the world's average temperature will make India's summer monsoon highly unpredictable.
- At 4°C warming, an extremely wet monsoon that currently has a chance of occurring only once in 100 years is projected to occur every 10 years by the end of the century.
- An abrupt change in the monsoon may precipitate a major crisis, triggering more frequent droughts as well as greater flooding in large parts of India.
- India's northwest coast and southeastern coastal region may see higher-than-average rainfall.
- Dry years are expected to be drier and wet years wetter.

To better understand the risks of climate change to development, the World Bank Group commissioned the Potsdam Institute for Climate Impact Research. This goal, however, as in the 2015 Paris Agreement on climate change, could turn into a fiasco without positive efforts being assured to reduce emissions of greenhouse gases (GHGs), for which developing countries are least responsible but the worst sufferers.

Impacts of climate change on socio-economic status and food security

The impact of climate change as a driver of future forced migration depends on several factors (IOM Report, 2008):

- The quantity of future greenhouse gas emissions;
- The rate of future population growth and distribution;

- The meteorological evolution of climate change;
- The effectiveness of local and national adaptation strategies.

Scientists are warning that if the present trend of 'global warming' continues, it will hamper agricultural production, and this in turn will affect the essential food supply. Climate change knows no borders. In South Asia, countries like India, Pakistan and Bangladesh have a profound potential threat of suffering from unhealthy high temperatures. At least 15% of them could be adversely affected. The recent severe drought of 2015–16, in which 330 million people were reported to have been affected, has provoked a growing awareness of this trend, with reports of mass migration to India's cities, leaving the rural villages empty (IPCC Report, 2014).

"India, taken as a whole, is experiencing warming as an integral part of the warming that is happening globally due to global climate change," said Robert S Ross of Florida State University, who conducted the study along with researchers from the Indian Institute of Technology, Bhubaneswar, and India Meteorological Department (IMD) (Sharma, 2018). Past records and evidence prove that climate change is a convincing truth. This phenomenon is currently far more rapid compared to the slow climate change that occurred in the past. Climate change often remains invisible or unrealised by the people. But with its further intensification when it happens at a rapid pace, it endangers the livelihoods of people over a region. All the countries in South Asia have become vulnerable in such a changed environment.

In the past, to deal with the changing situation, people used to move to rural areas, while the cities continued with reduced populations. Many small rural settlements located in the Himalayan foothills in the interfluves of the Ganga and Yamuna Rivers developed as a consequence.

The impacts of changing climate in the context of the economic situation of developing countries have already attracted the attention of the people and authorities concerned. A large part of the population in the region lives in poverty (Asian Development Bank estimates 903 million struggle on $1.25 per day) and their well-being is highly vulnerable to environmental hazards and degradation of natural resources (ADB Report, 2009). It is estimated that its impact on global economy is $60 trillion, most of which will be borne by developing countries with extreme weather, poor health conditions and low agricultural production (Moudgil, 2017). Warmer temperatures are already altering the timings of bird migrations, and perhaps climate change–induced human migration is not far behind. Since 2008, however, 161 countries have witnessed mass exoduses of people due to weather-related events. The United Nations High Commissioner for Refugees (UNHCR) says 36 million people were displaced by natural disasters in 2009. The number of environmental migrants in the world is estimated to rise between 25 million to 1 billion by 2050 (Kaur and Harpreet, 2018), the majority of whom will be from developing countries.

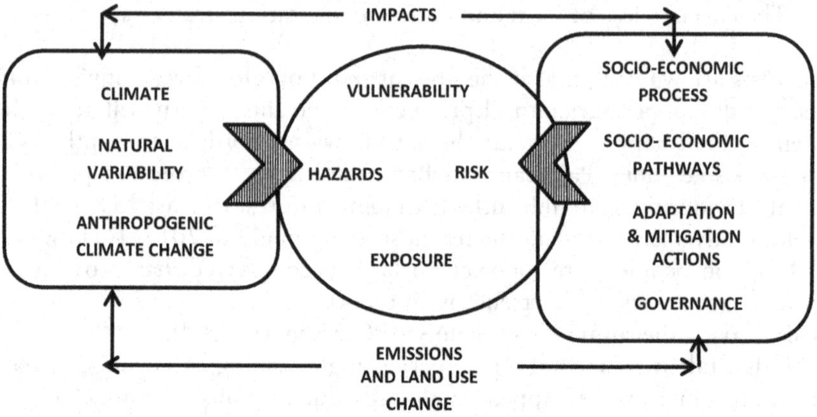

Figure 4.1 Multiple causes of climate changes and their impacts

Source: Flowchart prepared by the author based on IPCC Fifth Assessment Report, 2014

It has been assessed numerically and geographically that South and East Asia are particularly vulnerable to large-scale forced migration due to sea level rise alone, as a very dense population live in low-lying areas and a number of megacities are located on the coastline.

South Asia suffers from high levels of poverty, poor governance and projections of significant further population growth in all but Sri Lanka, with the population projected to more than double in Nepal and Pakistan. As such, it represents a major focus of global concern about climate change and migration (ADB report, 2009). In South Asia, governments, businesses and communities will have to take both short- and long-term approaches to manage climate risks. In the short term, integrating climate adaptation and disaster risk reduction will help them withstand shocks to human security and economic development from which recovery seems to be very costly (IPCC Report, 2014).

Emerging issues – crime, legalities and policy-making

Crime

As migrated people are displaced from their longtime habitation and forced to leave their assets and other belongings, a feel of deprivation inspires them to get involved in anti-social activities. Apart from these unfamiliar environments and other uncertainties regarding the availability of food, water and other basic needs, this kind of displacement exacerbates the tension that triggers conflicts. This could also lead to ethnic tensions that disrupt social harmony. Examples may be cited from the events when Bangladeshi immigrants seek employment in India or Afghan refugees in Pakistan. Sometimes the tense situation aggravates drug trafficking, smuggling and terrorist activities. South Asia is considered the second most violent place on earth after Iraq (Kaur and Harpreet, 2018). Climate-induced migration will worsen these existing tensions. The drought that preceded the rebellion in Syria is considered the worst long-term drought and the most severe set of crop failures since agro-based civilisation began in the region (Moudgil, 2017).

Legalities

Cross-border migration often raises questions on the legal status of the migrants. There is also no recognition of climate-induced migrants. This necessitates an official definition at the international level. There is a paucity of data on climate-induced migrants which prevents any government from framing policies in their favour, particularly at an international level. Maria Cristina Garcia, Department of History, Cornell University, in her book *Climate Refugees: The Environmental Origins of Refugee Migrations*, looked at environmentally driven migration as a more fast and furious historical phenomena. She also opined that the situation will require incredible adaptability and the political will to keep up with the changes (Swift, 2017). In Bangladesh, Soumya deep Banerjee, a migration specialist with the International Centre for Integrated Mountain Development (ICIMOD), said, "What we see here is a wide range of mobility outcomes – from evacuation and displacement to short- to long-term migration as adaptation, and planned resettlement."
ICIMOD has ranked India at the top amongst the countries that face displacement on account of sudden repetitive severe climatic changes in South and South East Asia. (The Third Pole, 2018).

Moreover, the term 'climate refugee' is controversial. In many countries, climate-induced migrants are difficult to recognise and are legally and politically left apart. These people are not afforded the same status as the mainstream population where they have migrated.

Societal response and adaptation

It is an ensuing problem in many places related to the interrelationship between migrants and the previous inhabitants. Migrants coming from a land with a different cultural setup more often face social rejection in many respects. Sometimes this culminates in living with separate identities with detached social perspectives. Politicians many times find the situation advantageous for them and, instead of bridging the gap, it is reinvigorated to linger. The mental conditions, psychological behaviour and adaptation capabilities in the social context of the migrants are very often overlooked by policy-makers. Concentrating on the economic aspects of migration has also shown it to be a striking problem, particularly in developing countries. However, in the Western context, climate change often remains an abstract concept, an issue that is "remote both in space and time; it is perceived as affecting other communities and future generations" (Lorenzoni and Pidgeon, 2006). Contrastingly, in India and in many other developing countries, weather extremes already have a significant and clearly visible detrimental effect on peoples' lives. Climate change will most probably exacerbate existing weather hazards, and it will add new and in some cases unknown risks (IPCC, 2007). Hence, the effect of climate change on migration decisions depends crucially on socio-economic, political and institutional conditions (Waldinger, 2015). As climate change and its victims have become a political issue, it has widely attracted the attention of mass media. The media's responsibility in a culturally and socially diversified country like India is immense. In fact, creating positive mass opinions and awareness could solve such societal and political problems (Reusswig et al., 2010).

Policy-making

At present, it is a challenge to policy-makers to set down a comprehensive plan for climate-induced migrants, though in some countries, it is a sensitive political issue. In India, the Nationally Determined Contributions (NDC) has stated its position on climate change and the frequent variability of extreme weather like droughts, floods, cyclones and so on and has also identified how rain-fed agriculture covers 60% of the country's total net sown area, with 40% of the total food production being affected by this. But it has remained silent about climate-induced migration. The policies of Bangladesh are also silent about the adaptation potential of migrants (The Third Pole, 2018). The South Asian Association for Regional Cooperation (SAARC) has noted and recognised climate change's consequences on the economy and society. In 2007, SAARC announced a declaration on climate change, and in 2008, a climate change action plan was adopted, including mitigation action for the responsible factors (Anderson et al., 2016).

Controversies over the terms used

The IPCC's Fifth Assessment Report conveyed that South Asia's climate is already changing, and further change is inevitable. The impacts are already being felt. Since the mid-19th century, the average increase in the temperature of the Earth's surface has been 0.85°C. Globally, sea levels have risen faster than at any time during the previous two millennia – and the effects are felt in South Asia (IPCC Report, 2014). Nevertheless, there is little evidence that climate change has directly influenced contemporary migration patterns, although historically, climate change has significantly impacted changing population distribution (ADB Report, 2009). At present, climate-induced migration has been termed environmental migration by many scientists. Migratory events are being termed migration dynamics and climate-induced migrants environmental refugees. But legally, the term 'refugee', as recognised in 1951 in the Geneva Convention Relating to the Status of Refugees (Refugee Convention), is absent. Legally speaking, the term 'environmental refugee' is a misnomer (Waldinger and Fankhauser, 2015). Waldinger and Fankhauser (2015) have justified by putting forward the reasons that it is difficult to isolate the environmental factors from other (often related) drivers of migration. Rather it stresses decisions taken by people confronted with gradual environmental change and focuses less on people forced to leave by sudden environmental shocks and natural disasters. Whatever the use of terms, the present perspective is to look at the historical features of climate change and human response initially to understand the essentials, but the focus is on managing the environmental situations in light of economic and political terms. In other words, the close nexus of all these terms together cannot be dissociated in further studies or research works. However, the core issues of the impact-creating factors remain the same in their uses. In future, it is presumed that there will be less human migration compared to what happened in historical times. This is because of the enhanced restrictions at international borders and effective international laws that may hinder mass migration.

Concluding remarks

Climate change is a complex environmental process and is not uniform everywhere. People's adaptability also differs. Adaptability is interlinked with socio-economic, political, cultural, technological, organisational, institutional or other variables which interact with it. People are not facing a uniform situation everywhere, and their access to natural resources varies. This incongruity leads to differences in adaptation approaches to the change of climate. The climatic situation is not the same everywhere. In a general sense, India gets an adequate monsoon, but the effects of the monsoon in the form of rainfall are extremely varied over time and space. Hence, the adverse effects of climate change are also not uniform over all political territories. Geographical factors are to be given priority, along with social, anthropogenic and other factors. Future policies, hence, are required to be region specific depending on the probability of the intensity of the vulnerability zones. Hence, a wide range of the number and nature of variables leads to regional variations of climate change–induced human migration. Peoples' perceptions should also be taken into account. At present, people, due to many political/legal status factors, cannot migrate freely and permanently as they used to do in ancient times. On the contrary, either they take refuge temporarily in some preferred safe place or try to overcome the situation of lower agricultural productivity and slower economic growth rate by adapting mitigation and specific risk management measures.

Climate change is a matter that ultimately indiscriminately affects all people in all respects. It is a hard fact, though, that poor agrarian communities are first affected. In places, it is only the displacement of people for temporary periods. People increase their adaptability with slow changes of climate until the vulnerability increases with severity. There are many synergies between action on climate adaptation and the achievement of growth and development objectives (IPCC Report, 2014). Climate change is clearly impacting South Asian communities' livelihoods and safety and increasingly forcing them to migrate (Anderson et al., 2016). New perspectives (area and action specific), objective characteristics, new research methodologies/strategies, improved technology and new adaptation strategies have emerged which might open up a better understanding of the definition of climate-induced migration and its socio-economic dimension in South Asia. Cooperation and integrated approaches at the regional, national and international levels are needed to achieve the best solution.

References

Anderson, Teresa, Shamsuddoha, Md, and Dixit, Ajaya. *Climate Change Knows No Border – An Analysis of Climate Induced Migration, Protection Gaps and Need for Solidarity in South Asia Analysis; Climate Change, Migration and South Asian Solidarity*. Dhaka, Bangladesh; BROT (Bread for the World)–Germany; ACTIONAID–Johannesburg, South Africa: CANSA (Climate Action Network South Asia), December 2016.

Cameron A. Petrie, et al. 'Adaptation to Variable Environments, Resilience to Climate Change, Investigating Land, Water and Settlement in Indus Northwest India.' *Current Anthropology: The University of Chicago Press Journals*, 58(1) (2017); Chicago, Il, USA.

Carling, Jorgen. *Migration, Human Smuggling and Trafficking from Nigeria to Europe*. IOM Migration Research Series, 23, Geneva, Switzerland: PRIO (The Peace Research Institute; Oslo), 2006.

Farrant, Macha, Anna MacDonald, and Dhananjayan Sriskandarajah. *Migration and Development: Opportunities and Challenges for Policymakers*. Ilse Pinto-Dobernig (Ed.), IOM Migration Research Series; No. 22, Geneva, Switzerland: International Organization for Migration, 2006.

IOM (International Organisation for Migration). *Migration and Climate Change*. IOM Migration Research Series No. 31, Geneva, Switzerland: IOM, 2008.

IPCC. 'Impacts, Adaptation and Vulnerability.' In M. L. Parry, O. F. Canziani, J. P. Palutikof, P. J. van der Linden, and C. E. Hanson (Eds.), *Contribution of Working Group II to the Fourth Assessment Report of the Intergovernmental Panel on Climate Change*. Cambridge, UK: Cambridge University Press, 2007.

IPCC (Intergovernmental Panel on Climate Change). *What's in It for South Asia?* Fifth Assessment Report, Overseas Development Institute and Climate and Development Knowledge Network, 2014.

Kathayat, Gayatri, et al. 'The Indian Monsoon Variability and Civilization Changes in the Indian Subcontinent.' *Science Advances, AAAS* (December 2017). Available at: https://www.ncbi.nlm.nih.gov/pmc/articles/PMC5733.

Kaur, Simrit, and Kaur, Harpreet. 'A Storm of Climate Change Migration Is Brewing in South Asia.' *East Asia Forum, Economics, Politics and Public Policy in East Asia and the Pacific*, New Delhi, June 17, 2018.

Khan, M.Z.A., and Sonal, Gangawala. *Global Climate Change – Causes and Consequences*. New Delhi: Rawat Publications, 2011.

Korisettar, Ravi. *Irfan Habib – Prehistory and Irfan Habib, The Indus Civilisation*. December, 2015. Available at: www.researchgate.net/publication/285543638.

Kumar, Vinod. 'Role of Indigenous Knowledge in Climate Change Adaptation Strategies: A Study with Special Reference to North-Western India.' *Journal of Geography & Natural Disasters* (November 10, 2014). ISSN: 2167–0587; Haryana.

Lorenzoni, I., and Pidgeon, N. 'Public Views on Climate Change: European and USA Perspectives.' *Climatic Change*, 77 (2006): 73–95.

Misra, V. N. 'Climate, a Factor in the Rise and Fall of the Indus Civilization – Evidence from Rajasthan and beyond, in Frontiers of the Indus Civilization' In *Sir Mortimer Wheeler Commemoration* (Vol. 143, pp. 461–490). Cambridge University Press, London; Reprint in India 1984.

Moudgil, Manu. 'Your Story.' In *How Climate Change Led to the Fall of Indus Valley Civilisation*. July 19, 2017. Available at: https://yourstory.com/2017/07/climate-change-indus-valley-civilisation-harappa/amp.

Piguet, Etienne, Pécoud, Antoine, and Guchteneire, Paul de. 'Migration and Climate Change: An Overview.' *Quarterly*, 33(3) Refugee Survey (2011): 1–23.

Rabbani, Golam, Shafeeqa, Fathimath, and Sharma, Sanjay. *Assessing the Climate Change Environmental Degradation and Migration Nexus in South Asia*. IOM Development Fund. Dhaka, Bangladesh: IOM (International Organisation for Migration), December 2016.

Report, Asian Development Bank (ADB). *Climate Change and Migration in Asia and the Pacific*. Executive Summary (Draft). Philippines: The University of Adelaide Flinders University, The University of Waikato, 2009.

Reusswig, Fritz, and Meyer-Ohlendorf, Lutz. *Social Representation on Climate Change: A Case Study from Hyderabad (India)*. Germany: Europäischer Hochschulverlag GmbH & Co. KG, January 2010.

Sarkar, A., Mukherjee, D., Bera, M., Das, B., Juyal, N., Morthekai, P., Deshpande, D., Shinde, R. S. V., and Rao, L.S.V. 'Oxygen Isotope in Archaeological Bioapatites from India: Implications to Climate Change and Decline of Bronze Age Harappan Civilization.' *Science Report*, 6 (2016): 26555.

Science Daily. Ancient Humans Left Africa to Escape Drying Climate; University of Arizona. October 4, 2017. Available at: www.sciencedaily.com/releases/2017/10/171004151231.htm.

Shaowu, Wang. 'Abrupt Climate Change and Collapse of Ancient Civilizations at 2200 BC–2000 BC.' *Progress in Natural Science; Taylor & Francis On line Journal*, 15(10) (2005): 908–914.

Sharma, C. Dinesh. 'India Has Warmed Rapidly in the Past 70 Years Study.' *Down to Earth*, Wednesday 16 May 2018.

Swift, Jackie. 'Migration, Forced by Climate Change.' Research, Department of History, Cornell University, 2017.

The Third Pole.Net. *South Asia Needs a Plan to Address Climate-Induced Migration; Climate-Induced-Migration-India-Bangladesh-Myanmar*. 2018. Available at: www.thethirdpole.net.

The World Bank Report. Feature Story, Toggle Navigation, June 19, 2013.

Tierney, Jessica E., Menocal, Peter B. de, and Zander, Paul D. 'A Climatic Context for the Out-of-Africa Migration.' *Geology* (2017). https://doi.org/10.1130/G39457.

Waldinger, Maria. 'The Effect of Climate Change on Internal and International Migration: Implications on Developing Countries.' Working Paper No. 217 & 192, London; UK: Centre for Climate Change Economics and Policy & Grantham Research Institute on Climate Change and The Environment, 2015.

Waldinger, Maria, and Fankhauser, Sam. *Climate Change and Migration in Developing Countries: Evidence and Implications for PRISE Countries*. London: ESRC Centre for Climate Change Economics and Policy and Grantham Research Institute on Climate Change and the Environment, October 2015.

Worland, Justin. 'Climate Change Will Make Parts of South Asia Unlivable by 2100, Study Says.' *TIME*, August 2, 2017. Available at: https://time.com/4884648/climate-change-india-temperatures/ (Accessed on 16 December 2019).

5 Impacts of climate change on migration and economic growth

South Asian perspectives

Moushakhi Ray

Today, nations are satisfied not simply with economic development but with sustainable economic development. The Brundtland Commission Report (1987) introduced sustainable development as a process that "meets the needs of the present without compromising the ability of future generations to meet their own needs."

This chapter considers that the economic progress of a country is tilted primarily towards the growth of its manufacturing sector, including heavy industries, which might cause environmental degradation. Indeed, growth behaviour, as observed in the past for present-day developed nations, motivates the South Asian countries (notably, Afghanistan, Bangladesh, Bhutan, India, Maldives, Nepal and Pakistan) to prioritise the development of their manufacturing sectors. In this journey of economic growth, concern for the environment goes largely unnoticed. Given this background, this chapter aims to study the interrelationship among economic growth, migration and climate change in South Asian countries. It uses a theoretical model to argue that higher levels of economic growth and export exert damaging impacts on climate (as represented by the level of pollution), and a higher amount of emissions (that is, pollution) leads to higher productive activities. To be more specific in this regard, a higher level of capital, or a larger volume of labour in the production process, will lead to a deterioration of the climatic conditions. It further argues that immigration has a positive impact on economic growth and the level of pollution. The theoretical underpinnings in the relationship among climate change, migration and economic growth are largely corroborated by empirical results for South Asian countries.

The chapter is structured as follows: the second section deliberates in a nutshell the interrelationship among climate change, manufacturing and economic growth and global views in those respects. The third section briefly discusses various empirical and theoretical studies relating to sustainable development. The fourth section develops a theoretical model incorporating Cobb-Douglas production and welfare functions, based on Lopez (1994)'s theoretical underpinnings. The theoretical findings are tested empirically for South Asian countries in the fifth section, and the conclusion is presented in the sixth.

Relationship among growth, manufacturing and pollution: global views

While discussing the importance of manufacturing activities in economic growth, the stalwart economist Prof. Kaldor has advocated three laws. Those are: (a) the growth of GDP is positively related to the growth of the manufacturing sector, (b) the productivity of the manufacturing sector is positively related to the growth of the manufacturing sector (also known as Verdoorn's law) and (c) the productivity of the non-manufacturing sector is positively related to the growth of the manufacturing sector. These Kaldorian hypotheses have been empirically tested in a number of studies across a range of developing countries, including South Africa (see, for example, Dasgupta and Singh, 2005b; Millin and Tenassie, 2005; Wells and Thirlwal, 2004), where manufacturing displays a strong positive correlation with GDP growth. Among the developed economies in 2005, according to the OECD, the manufacturing sector contributed 20.2% to GDP in Japan, 23.2% in Germany and 13.4% in the United States. Also, the Economic Policy Institute confirms that each manufacturing job supports almost three other jobs in the economy, suggesting that a rapid development of manufacturing activities in an economy redresses the unemployment problem to a good extent.

These theoretical underpinnings are corroborated in the development stories of present-day developed countries. The experiences of the US economy are in particular suggestive to this end. The share of the manufacturing sector in the US economy started increasing from the late nineteenth century, and, during the World War II, it became the main supplier of war materials. Its automobile sector, in particular, was transformed into a powerhouse for the production of such materials, which substantially augmented its overall productivity and output. In fact, the period 1940–44 represents the greatest expansion of industrial production that the United States had ever experienced. Notwithstanding its secondary sector's declining percentage in GDP since 1960s, the United States remains the world's largest manufacturer of goods. In 2004, its manufacturers turned over 70% higher than what Japan, the world's second largest manufacturer, did and over three times more than China. In fact, the United States produces about one-quarter of the global manufactured goods.

Such development stories of presently developed nations have motivated South Asian countries to adopt the path of manufacturing-led economic growth. Insofar as India is concerned, it is found that the economic reforms of the early 1990s placed much weight on the manufacturing sector rather than agriculture. Indeed, the growth performance in South Asia is impressive in recent years. In the 1990s, the average annual GDP growth in this region ranged from 4% in Pakistan to 5.5% in India. These rates were relatively higher than the historical rates of growth in these countries and contributed much to the alleviation of poverty. The following six years,

2001–2006, further strengthened the trends of the 1990s: India's growth in this period rose by 2% to 7.5%; Pakistan grew by more than 5.5%; and Bhutan accelerated its growth from 5% to nearly 8%. Bangladesh also consolidated its growth performances. But in Nepal, the rate of economic growth retarded from 5% in 1990 to 3.5% and in Sri Lanka from more than 5% in 1990 to 4.5% in 2006, possibly as a result of insurgencies in those countries. In 2006, the rate of growth rose sharply to 9% in India, to 8% in Bhutan, to more than 7% in Pakistan and Sri Lanka, and close to 7% in Bangladesh. Also, the Global Competitiveness Report 2016–17 shows that most of the South Asian countries have improved their competitiveness over last few years. But a relative dark side of these economies is their labour markets, which have been deteriorating over the last 10 years. It concerns twin aspects simultaneously, the level of employment and the formation of technical skills.

There is no doubt that manufacturing-led economic growth helps reduce the incidence of unemployment, but very frequently, such growth processes are accompanied by various types of pollution. Indeed, a reduction in fresh air due to higher carbon dioxide emissions and burning fossil fuels are the consequences of the lack of the idea of sustainability in economic progress. Thus, in India, the carbon dioxide emission metric tons per capita was 0.71 in 1990 but rose to 1.72 in 2014 and 1.82 in 2016. In contrast, in the United States, it fell from 19.32 in 1990 to 16.42 in 2014 and then to 15.50 in 2016. These climatic features in South Asian countries are often explained by the essentialities of developmental requirements, which certainly ignore the question of sustainable development.

The question of greenhouse gases is of particular importance in this regard. Over the last 150 years, air pollution – especially fossil-fuel combustions – has been steadily increasing greenhouse gas concentrations in this planet, and the United States is the largest source of it. Since 1990, gross US greenhouse gas emissions have increased by about 2%. Only from 2016 onwards, such emissions from that source have been showing a declining trend, as less fossil fuel combustion has been taking place owing to the substitution of natural gas for coal in the power sector. Also, greenhouse gas emissions from US industries have declined by almost 14% since 1990. The reduction in carbon dioxide emissions and air pollution in developed countries like the United States is indicative of a gradual shift in their focus from manufacturing to the service sector. In order to ensure sustainable development, these countries have also been increasingly relying on several pollution-control measures in the manufacturing sector. But such a development paradigm has failed to garner wide support in South Asian countries thanks to their financial inabilities.

Against this background, the United Nations Commission on Sustainable Development (CSD) was established by the UN General Assembly in December 1992 to ensure effective follow-up of United Nations Conference on Environment and Development (UNCED), commonly known as the

Rio Conference or Earth Summit, which underpins public awareness about the need to integrate environment and development. In this connection, the UNCED recognises the need for substantial mobilisation of resources to provide strong financial support to developing countries in their efforts to promote sustainable development. It has influenced subsequent UN conferences, including setting out their global green agenda. The World Conference on Human Rights, for example, focuses on the right of people to a healthy environment and the right to development. Major outcomes of the conference include the United Nations Framework Convention on Climate Change (UNFCCC), which has led to the Kyoto Protocol, Agenda 21, the United Nations Convention on Biological Diversity (CBD) and the United Nations Convention to Combat Desertification (UNCCD). Climate change and its consequences were, however, widely recognised in 1988 when the Intergovernmental Panel on Climate Change (IPCC) was set up. Its findings highlight how there has been a marked increase in the global atmospheric concentrations of carbon dioxide, methane and nitrous oxide as a result of human activities since 1750.

Also, governments in some South Asian countries keep their eyes on the environment while implementing development projects. Thus, for example, China promotes clean, green and low-carbon cities through its International Cooperation Project. Also, China's Liaoning Safe and Sustainable Urban Water Supply Project aims to improve water quality and operational efficiency of selected water supply utilities in the project areas. In the Maldives, the Clean Energy for Climate Mitigation (CECM) Project is designed to reduce dependence on imported fossil fuels for power generation through the use of renewable energy resources and adoption of energy efficiency measures. Similarly, in India, the development objective of the Maharashtra Project on Climate Resilient Agriculture Project is to enhance climate resilience and the profitability of smallholder farming systems in selected districts of Maharashtra.

Certain private initiatives should also be mentioned in this context. In the United States, Swire Properties Inc. has developed various residential and commercial properties keeping in view the pollution aspects. The sustainable development policy of SWIRE – a wholly owned subsidiary of Swire Properties Inc. – gives priority to the environment in its business decisions.

Empirical and theoretical studies on sustainable development

A number of studies have been undertaken to highlight the significance of sustainable development. Kumar et al. (2010) report the proceedings of a workshop that has examined India's need to incorporate the issue of climate changes in various development policies. It stresses that climate change would have significant adverse impacts on agriculture, especially in developing countries. Given a large proportion of its population depending directly and indirectly on agriculture, such adverse effects would surely be reflected in the escalation of poverty. Agarwal (2009) notes that a one-degree rise in temperature might reduce the yields of wheat, soya bean, mustard, groundnut and potato by 3–7%. Again, Kumar and Parikh (2001) and Sanghi and Mendelsohn (2008) estimate that, under a scenario of moderate climate changes, there could be about a 9% decline in farm-level net revenue in India. For South Asia as a whole, Nelson et al. (2009) indicate that the daily per capita calorie availability would decline by about 8% in 2050. However, in October 2006, Stern noted that

> if we don't act, the overall costs and risks of climate change will be equivalent to losing at least 5 per cent of global gross domestic product (GDP) each year, now and forever. If a wider range of risks and impacts is taken into account, the estimates of damage could rise to 20 per cent of GDP or more.

There is, however, a methodological weakness in such studies. Citing the presence of strong spatial autocorrelation in the agricultural output data in India, Kumar (2009) argues in favour of controlling for spatial effects in the estimation of climate change impacts.

In recent years, there has been growing scholarship on the relationship between migration and environment. Richmond (1993) argues that more descriptive typologies of environmentally induced migration should understand the dynamic interaction of the multiple factors that generate such movement. Migration is often a logical response to environmental disaster, but it is rarely a solution for environmental problems. Environmental stability can only be attained by relatively lower levels of population growth and ecologically sustainable ways of using the environment. There has been a trend toward increasing numbers of people displaced by environmental disasters in Asia over the last decades. China and India (with 37.8% of the world's population) dominate the cases of population displacement. Whereas out-migration reduces the availability of workforce in South Asian countries, thereby bringing in higher remittances, the process of in-migration contributes to severe climatic changes. Intra-country migration is also important from the viewpoint of the environment. Presently, about 30% of the Indian population live in urban areas. By 2030, the number is expected to rise to around 40.6%, and, by 2050, India's urban population might even surpass the population living in rural areas (United Nations, 2007). Considering that Indian cities consume more fossil fuels than rural areas, this urbanisation process would drive up emissions at an exponential rate (McKinsey and Company, 2009).

A number of theoretical studies have also been undertaken on the interactions among population, migration, economic growth, environment, wealth creation and sustainability. In this context, Lopez (1994) and Anderson (2019) should be cited, who address various issues relating to the relationship between environmental degradation and economic growth. The issue of optimal allocation has also evolved in the discussion. Daly (1990) interprets the optimal allocation of resources in terms of labour, capital as well as natural resources. But natural resources are not considered a component in an entropic metabolic flow from and to the environment. Rather, they are indestructible building blocks in the circular flow. A Pareto optimal allocation can be achieved for any scale of population and per capita resource use. The concept of economic efficiency is indifferent to the scale of the economy's physical dimensions, just as it is indifferent to the distribution of income. Equity of income distribution and sustainability of scale lie outside the concept of market efficiency. But the environment is sensitive to the physical scale of the economy, and human welfare is sensitive to the drive of the environment. Optimal allocation of resources at a non-optimal scale requires selecting the best out of a number of bad situations. If the economy continues to grow beyond the environmentally sustainable optimal scale, then optimal allocation means to choose the best of an ever-worsening situation. This anomaly is absent from the circular flow vision.

Along a similar line, Beckerman (1994), King and Slesser (1995), Hecht (1999) and Diesendorf (2000) discuss sustainability and sustainable development. Their basic logic is this: the capacity of developing countries to attain sustainable development is affected by a number of achievable factors, which include (a) domestic policy actions, including measures for open and free market economies, sound and enforceable environmental policies and public participation in decision-making; (b) financing environmental policies of the bilateral and multilateral lending institutions; and (c) monitoring private-sector investment for the sake of clean technology. The interaction of these factors forms a triad for sustainable development. Hecht (1999) discusses these factors, suggesting supplementing a positive relationship among them in promoting sustainable development. Dua and Esty (1997), however, observe that, although some developing countries have adopted these measures, their implementations beg questions. This is quite evident in South Asia.

However, King and Slesser (1995) and Diesendorf (2000) conceive that sustainability and sustainable development are about establishing a rate of economic growth that is compatible with the rate at which the environment is able (a) to assimilate the waste products of human activities and (b) to provide the means of life support. That is, the path of economic growth must account for environmental stability, and, at the same time, the rate of wealth creation must be sufficient to sustain the economy and provide the necessary investments required for environmental and resource management. The trade-off between environmental stability and economic growth is studied there to confirm the interrelationship among population, economic growth and environment. They treat sustainability as the goal of a process called 'sustainable development' or 'ecologically sustainable and socially just development' (ESD).

According to Stock et al. (2017), however, global development is represented by an increasing socio-economic inequality, migration and displacement, urbanisation, climate change and environmental degradation, which direct the global community along a sustainably irresponsible development path. He describes the Fuzy Front End of Innovation (FFE) process,[1] which starts with an initiative based on creative observation and idea initiation. The impulse for an idea comes from the deviation between the initiator's expectations about a state and the observed state. This arouses the intention to close the observation gap and generates solution ideas. Subsequently, the FFE process is completed by assessing and selecting the best solution idea. The overall process is based on an iterative process. The sustainable invention must be subsequently transformed into a competitive innovation by the integrated development of a product and business model in an environment of extreme uncertainty. In line with this idea, Stock et al. (2017) represents the customer solution by a detailed business model with a related and fully designed and tested product. According to him, by means of market dynamics of cooperation and competition in global value creation and knowledge networks, the customer solution, based on a sustainable idea, can be a necessary driver for moving towards global sustainable development.

Subsequent to the literature on optimal allocation, several studies analyse the consequences of the inability to attain sustainable development in the process of prioritising economic growth, especially in South Asian countries. Obviously, a higher level of manufacturing growth without due emphasis on pollution control degrades the environment and thereby results in a lower rate of future economic growth and also a higher rate of out-migration. In connection with this, Padukone (2010) points out that economic opportunities in rural areas are meagre, which deters further investment in those places and compels people to migrate to urban areas in search of employment. A greater population in urban areas drives more investment, thereby perpetuating the flow of migration. He also validates that climate changes hamper agriculture, the availability of food and the creation of employment and also limit the productivity of workers in South Asian countries, thereby retarding their economic growth and inducing out-migration.

Hugo (1996), however, focuses upon international migration and its implications with respect to environmental changes. He considers international migration as an argument causing changes in the environment, using an equation as developed in Ehrlich and Holdren (1971). The finding of his mathematical model is that international migration plays a significant role in determining population size in the areas of origin and destination of migrants. With respect to cross-country migration, Padukone (2010) asserts that, with a potentiality to radically alter India's economy and to reshape the human security terrain and even the geopolitical order of South Asia, climate change is an issue where India perhaps is potentially the greatest loser. Adaptation to inevitable environmental shifts is imperative to manage the political and security-oriented consequences of climate change.

Regarding the policy perspective, however, Cooper (2008) points out that a carbon tax might be easily extended to reduce carbon dioxide emissions. Cooper (2008) and Nordhaus (2007) advocate for a uniform carbon tax being introduced across all countries. But one should recognise that the currently developed countries have contributed much to earlier GHG emissions and therefore to the current state of climatic disorder. Therefore, a uniform carbon tax, both for developed and developing countries, is never acceptable to India and other developing countries. Against this background, Dutt and Gaioli (2007) propose that the advantages of emissions trading and carbon tax should be combined in a system where emissions trading applies to transactions between industrialised and developing countries.

The most widely used economic models of climate change are integrated assessment models that link climate and economic simulations (Nordhaus and Yang 1996, 2007). These models start with the standard economic assumptions of rationality, perfect competition and optimising behaviour. The differences among the major climate models lie in assumptions regarding the rate of discounting climate change mitigation (that is, avoiding the costs of future climate damage to economic activity) and the costs of mitigation efforts. The standard formula that is used in these models is based on the works of Ramsey (1928), Arrow (1966) and Fellner (1967). However, these economic models of human development and climate change have been examined by Gowdy and Salman (2008).

Theoretical framework

The theoretical underpinnings for the empirical exercise in this chapter are based on Lopez (1994). The model considers a neoclassical-type production function

$$Y = G\,(f\,(K, L), x) \tag{5.1}$$

in the absence of technological indices. Here, *K* represents capital, *L* labour and *x* the environment factor (representing, for example, air emissions). In that case, the revenue function is

$$R = R\,[P; f\,(K, L), x] \tag{5.2}$$

where *P* is assumed to be a constant in a small open economy. The social welfare function is defined as

$$U = U\,[R\,[P; f\,(K, L), x], x] \tag{5.3}$$

If the Cobb-Douglas functional specification is taken into account, the production function is

$$Y = AK^{\alpha}L^{\beta}x^{\gamma} \tag{5.4}$$

and the revenue function

$$R = P\,AK^{\alpha}L^{\beta}x^{\gamma} \tag{5.5}$$

The Cobb-Douglas welfare function with the consumer good *Y* and emission x is considered:

$$U = B\,Y^{\tau}\,x^{-\omega} = BA^{\tau}K^{\alpha\tau}L^{\beta\tau}x^{\gamma\tau-\omega} = C\,K^{\alpha\tau}L^{\beta\tau}x^{\gamma\tau-\omega} \tag{5.6}$$

where $C = BA^{\tau}$.

For the utility maximisation under the general functional forms, we have from Equation (5.3):

$$\frac{\partial U}{\partial x} = \frac{\partial U}{\partial R}.\frac{\partial R}{\partial x} + \frac{\partial U}{\partial x} = 0$$

$$\frac{\partial R}{\partial x} = R_2(.) = -\frac{\frac{\partial U}{\partial x}}{\frac{\partial U}{\partial R}} = -\frac{U_2(.)}{U_1(.)} \tag{5.7}$$

where $U_1(.)$ is the marginal utility of the consumer good, and $U_2(.)$ is that of emission.

For the Cobb-Douglas welfare function (5.6), we obtain

$$U_1(.) = \frac{\partial U}{\partial R} = \frac{\partial U}{\partial Y}$$

standardising the price level at unity. Therefore,

$$U_1(.) = \frac{\partial U}{\partial Y} = C_1\tau\left(K^{\alpha}L^{\beta}x^{\gamma}\right)^{\tau-1}x^{-\omega} = C_1\tau K^{\alpha(\tau-1)}L^{\beta(\tau-1)}x^{\gamma(\tau-1)-\omega}$$

$$= C_1\tau K^{\alpha\tau-\alpha}L^{\beta\tau-\beta}x^{\gamma\tau-\gamma-\omega} \tag{5.8}$$

where $C_1 = BA^{\tau-1}$. Again,

$$U_2(.) = \frac{\partial U}{\partial x} = C(\tau\gamma-\omega)K^{\alpha\tau}L^{\beta\tau}x^{\gamma\tau-\omega-1} \tag{5.9}$$

Hence,

$$R_2(.) = -\frac{C(\tau\gamma-\omega)K^{\alpha\tau}L^{\beta\tau}x^{\gamma\tau-\omega-1}}{C_1\tau K^{\alpha\tau-\alpha}L^{\beta\tau-\beta}x^{\gamma\tau-\gamma-\omega}}$$

$$= -\frac{A(\tau\gamma-\omega)}{\tau}K^{\alpha}L^{\beta}x^{\gamma-1}$$

$$= -\left(\gamma - \frac{\omega}{\tau}\right)\frac{Y}{x} \tag{5.10}$$

Here, γ represents the effects of environmental factor on production, and $\frac{\omega}{\tau}$ is the effect of environmental factor on welfare per unit of its effect on the production of consumer good. If the latter effect is greater than the former,

$$\frac{\partial R}{\partial x} = \frac{\partial Y}{\partial x} = R_2(.) > 0 \tag{5.11}$$

That is, a higher amount of emission leads to higher productive activities.

Returning to Equation (5.7), it can be noted that

$$R_2(.) = -\frac{U_2(.)}{U_1(.)} = -\frac{\frac{\partial U}{\partial x}}{\frac{\partial U}{\partial R}} = \frac{\partial R}{\partial x} = \frac{\partial Y}{\partial x} \tag{5.12}$$

That is, $R_2(.)$ is the marginal revenue product of the environmental factor. Now, if q is considered a constant (that is, it does not reflect the changes in consumers' willingness to accept pollution), a profit-maximising producer would equate it with the marginal cost of environmental factor, which is the price of emission (say, q). However, an existence of a price of emission is treated as a pollution control measure. Hence,

$$R_2(.) = -\frac{U_2(.)}{U_1(.)} = q \tag{5.13}$$

Totally differentiating Equation (5.13) with respect to x and Y, we get

$$R_{21}dY + R_{22}dx = 0 \tag{5.14}$$

where $R_{21} = \frac{\partial R_2}{\partial Y}$ and $R_{22} = \frac{\partial R_2}{\partial x}$. Hence

$$\frac{dx}{dY} = -\frac{R_{21}}{R_{22}} \tag{5.15}$$

Differentiating Equation (5.10) with respect to Y and x, we get

$$R_{21} = \frac{dR_2}{dY} = -\left(\gamma - \frac{\omega}{\tau}\right)\frac{Y}{x} \tag{5.16}$$

$$R_{22} = \frac{dR_2}{dx} = \left(\gamma - \frac{\omega}{\tau}\right)\frac{1}{x^2} \tag{5.17}$$

respectively. Therefore,

$$\frac{dx}{dY} = -\frac{-\left(\gamma - \frac{\omega}{\tau}\right)\frac{Y}{x}}{\left(\gamma - \frac{\omega}{\tau}\right)\frac{1}{x^2}}$$

$$= \frac{Y}{X} \tag{5.18}$$

which is positive. Thus, a higher amount of capital or a larger volume of labour (caused by the natural growth of population or a higher migration) would lead to deterioration of the environment.

Empirical findings

A few studies are carried out on the relationship among economic growth, trade liberalisation and the degradation of environment. But, as Lopez (1994) observes, "In general, there is very little agreement on the nature of the linkages among trade reform, growth, and environmental degradation" (p. 163).

We seek to estimate here such a relationship for South Asian countries. Our model in this case is

$$\text{Em} = \beta_0 \text{EcG}^{\beta 1} \text{Exp}^{\beta 2} \text{Dbt}^{\beta 3} \text{U} \tag{5.19}$$

Where Em represents CO_2 emissions (expressed in metric tons per capita per year), EcG is economic growth (expressed in percentage per annum), Exp is exports (expressed in percentage of GDP) and Dbt is the external debt (expressed in percentage of GNI). The external debt is used as a control variable to avoid its effects on environmental degradation. The reason for controlling this variable is that, in many cases, external debt is tied with the import of plants and machineries (with embedded technology), which might have impacts on environment in any direction.

The relationship is estimated using panel data where the time series run from 1998 to 2014 for six South Asian countries: Bangladesh, Bhutan, India, Maldives, Nepal and Pakistan. In the first place, the Hausman test is carried out using the software Stata to examine the suitability between the fixed-effect and random-effect models. The test shows the *p*-value $>\chi^2$ is 0.66 so that at 0.01 level, we reject the null hypothesis of the suitability of the fixed-effect model and proceed to estimate the random-effect model. In the raw data, both autocorrelation and heteroskedasticity prevail; those are duly corrected in the final dataset. The Wald $\chi^2(4)$ statistic is found at 4.92, whose p-value is greater than 0.2961.[2] The significance level is thus on the lower side. However, the estimated model is as follows:

$$\begin{aligned} Em =& -0.0767 + 0\ 0222\ \ln EcG + 0.0799 \ln Exp - 0.1658 \ln Dbt + e \quad (5.20) \\ & (z = -0.14) \quad (z = 1.24) \quad (z = 0.84) \quad (z = -1.57) \\ & (P>z = 0.887)\ (P>z = 0.216)\ (P>z = 0.402)\ (P>z = 0.116) \end{aligned}$$

From this estimated relationship, three inferences are suggestive. One, the variable EcG is significant at about 80% and the variable Exp at 60%, while the control variable is found significant at more than 90%. Two, the signs of the estimated coefficients indicate the variables EcG and Exp exert damaging impacts on environment, but external debt is expected to help reduce CO_2 emissions. The promotion of manufacturing-led economic growth was initially practiced by developed countries. In line with the history of higher growth rates in developed countries, the rise in the rate of economic growth necessarily indicated progress in the secondary sector of the South Asian countries. These are developing countries lacking in sufficient investment in pollution control measures. The increase in production of manufacturing sectors contributes to raising the level of pollution and thereby degrades climatic conditions. Therefore, unlike developed countries with established pollution control measures, the direct relationship between economic growth and degradation of the environment (measured by CO_2 emissions) is prevalent in South Asian countries. Also, a higher export level increases revenue generation and foreign exchange reserve of the economy and contributes to economic growth, thereby rendering a positive relationship with the degradation of the environment. Therefore, promotion of export-led manufacturing growth with policy initiatives to control pollution in South Asian countries is a necessity to climb the ladder of sustainable economic development. The latter inference on the inverse relationship between external debt and CO_2 emission may possibly be explained by the fact that while utilising external debt for development purposes, South Asian countries pay attention to environmental issues in decision-making. Three, the values of estimated coefficients suggest that a 1% increase in economic growth would lead to a 2.22% rise in CO_2 emissions and that a 1% increase in exports would cause about an 8% rise in emissions. Perhaps South Asian countries' higher proportion of manufacturing products in exports, relative to its proportion in their total economic activities, explains why their exports contribute more to pollution than does their economic growth.

To further investigate the impacts of economic growth on pollution, we hypothesise here that immigration generates higher economic growth in a country, which, as we have found previously, leads to a higher level of pollution. Immigration increases population pressure in South Asian countries and directly contributes to greater pollution. Increased population increases the availability of the work force and the production level in these economies. Higher production contributes to higher economic growth, which is necessary to sustain the livelihood of immigrants. Higher growth increases the level of pollution. On these hypotheses, we consider the following models:

$$\text{EcG} = \beta_0 + \beta_1 \text{Mig} \tag{5.21}$$

$$\text{Em} = \delta_0 + \delta_1 \text{Mig} \tag{5.22}$$

where Mig represents the number of immigration (number of people entering the country). Using the same dataset underlying the estimation in Equation (5.19), we compute the models in Equation (5.21) and (5.22). There is neither any autocorrelation nor a heteroskedasticity problem in these estimations. The estimated models are presented subsequently.

$$\begin{aligned} \text{EcG} = {} & 6.8337 && + \quad (5.36\text{e} - 07)\ \text{Mig} \\ & (z = 7.99) && \quad\ (z = 0.86) \\ & (p{>}z = 0.000) && \quad\ (p{>}z = 0.392) \end{aligned} \tag{5.23}$$

$$\begin{aligned} \text{Em} = {} & 0.8228 && + \quad (4.80\text{e} - 07)\ \text{Mig} \\ & (z = 4.88) && \quad\ (z = 0.38) \\ & (p > z = 0.000) && \quad\ (p > z = 0.704) \end{aligned} \tag{5.24}$$

The estimated models (5.23) and (5.24) suggest that the variable immigration assumes a positive coefficient in the explanation of both economic growth and emission, which is our *a priori* argument.[3] But, while the immigration coefficient is significant at about the 61% level in the explanation of economic growth, it is not significantly different from zero in the explanation of emissions.

Concluding remarks

This chapter thus highlights that there are a number of studies indicating the growing importance of manufacturing activities in path of economic growth, but such a growth pattern leads to a higher level of pollution. The literature indeed supports the view that there is close relationship between economic growth and the level of pollution and also between trade liberalisation and pollution. Following the theoretical layout of Lopez, we have used a Cobb-Douglas model to study the effects of economic growth on pollution and also the effects of pollution on economic growth. The chapter reveals that higher levels of economic growth and exports increase the level of pollution and degrade climatic conditions. Empirically, the relationship between pollution, on the one hand, and economic growth and exports, on the other, has been tested for South Asian countries. The *a priori* relationship is corroborated in the empirical exercise. This chapter also verified the relationship between economic growth and immigration and between the level of pollution and immigration. An increase in immigration causes higher pollution, and the increased labour force enhances the productive activity of the economy.[4]

Against this background, it is crucial to strike a balance between economic growth and the intended level of pollution. Promotion of export-led manufacturing growth with policy initiatives to control pollution in South Asian countries is a necessity to climb the ladder of sustainable economic development. To minimise the adverse effects of pollution, steps should be taken in the field of strategic investments in research and development for environment-friendly technological innovations, capital expenditure (CAPEX) on actual energy projects and greenhouse gas (GHG) emission trading. We note in this context that the SmartWay Program adopted by the US Environmental Protection Agency (EPA) has empowered companies to move goods in the cleanest and most energy-efficient way possible. Similarly, the California Sustainable Freight Action Plan sets a goal of using zero or near-zero emissions equipment to transport freight everywhere feasible. Also, governments should impose an optimal tax at the firm level to trade off between environment hazards and growth prospects (Gielen and Moriguchi, 2002). In a similar tone, this chapter identifies the existence of a price for pollution as a pollution control measure.

Notes

1 FFE (fuzzy front end innovation) is defined as those activities that come before formal and well-structured new product and process development. The FFE process is the first prerequisite for any innovation and addresses the generation of ideas.
2 See Table 5.1-A.
3 See Table 5.2-A.
4 See Tables 5.3-A, 5.4-A and 5.5-A.

References

Agarwal, Bina. 'Gender and Forest Conservation: The Impact of Women's Participation in Community Forest Governance.' *Ecological Economics*, 68(11) (2009): 2785–2799.

Anderson, Dennis. 'Economic Growth and the Environment.' Policy Research Working Papers, Background paper for World Development Report, 1992. Available at: http://documents.worldbank.org/curated/en/509711468739185650/Economic-growth-and-the-environment (Accessed on 13 January 2019).

Arrow, Kenneth J. 'Discounting and Public Investment Criteria.' In A.V. Kneese and S. C. Smith (Eds.), *Water Research* (pp. 13–32). Baltimore, MD: Johns Hopkins University Press, 1966.

Beckerman, Wilfred. 'Sustainable Development: Is It a Useful Concept?' *Environmental Values*, (3) (1994): 191–209.

Cooper, N. Richard. 'The Case for Charges on Greenhouse Gas Emissions.' Discussion Paper 08–10, *The Harvard Project on International Climate Agreements*, Department of Economics, Harvard University, October 2008.

Daly, E. Herman. 'Sustainable Development: From Concept and Theory to Operational Principles.' *Population and Development Review,* 16 (1990): 25.43.

Dasgupta, Sukti, and Singh, Ajit. 'Will Services Be the New Engine of Economic Growth in India?' Working Paper No. 310, Centre for Business Research Programme on Enterprise and Innovation, and Faculty of Economics, University of Cambridge, September 2005a. Available at: www.cbr.cam.ac.uk.

Dasgupta, Sukti, and Singh, Ajit. 'Manufacturing, Services and Premature De-Industrialisation in Developing Countries: A Kaldorian Empirical Analysis.' Working Paper No. 327, Centre for Business Research Programme on Enterprise and Innovation, and Faculty of Economics, University of Cambridge, June 2005b. Available at: www.cbr.cam.ac.uk.

Diesendorf, Mark. 'Sustainability and Sustainable Development.' In Dexter Dunphy, Jodie Benveniste D, Andrew Griffiths, and Philip Sutton (Eds.), *Sustainability: The Corporate Challenge of the 21st Century* (pp. 19–37). Sydney: Allen & Unwin, 2000.

Dua, Andre, and Esty, Daniel C. *Sustaining the Asia Pacific Miracle: Environmental Protection and Economic Integration.* New York: Columbia University Press, 1997.

Dutt, Gautam, and Gaioli, Fabian. 'Coping with Climate Change.' *Economic and Political Weekly,* 42(42) (2007): 4239–4250.

Ehrlich, R. Paul, and Holdren, John P. 'Impact of Population Growth.' *Science New Series,* 171(3977) (1971): 1212–1217.

Fellner, William. 'Operational Utility: The Theoretical Background and Measurement.' In William Fellner (Ed.), *Ten Economics Studies in the Tradition of Irving Fisher* (pp. 39–74). New York: John Wiley & Sons, 1967.

Gielen, Dolf, and Yuichi, Moriguchi. 'CO_2 in the Iron and Steel Industry: An Analysis of Japanese Emission Reduction Potentials.' *Energy Policy,* 30(10) (2002): 849–863.

Gowdy, John, and Salman, Aneel. 'Climate Change and Economic Development: A Pragmatic Approach.' *The Pakistan Development Review,* 46(4) (2008): 337–350.

Hecht, D. Alan. 'The Triad of Sustainable Development: Promoting Sustainable Development in Developing Countries.' *The Journal of Environment & Development,* 8(2) (1999): 111–132.

Hugo, Graeme. 'Environmental Concerns and International Migration.' *The International Migration Review,* 30(1) (1996): 105–131.

King, Jane, and Slesser, Malcolm. 'Prospects for Sustainable Development: The Significance of Population Growth.' *Population and Environment,* 16(6) (1995): 487–505.

Kumar, K.S. ' Climate Sensitivity of Indian Agriculture.' Working Papers No. 2009-043, Madras School of Economics, Chennai, India, 2009.

Kumar, Kavi K.S., and Parikh, Jyoti. 'Indian Agriculture and Climate Sensitivity.' *Global Environmental Change,* 1(2) (2001): 147–154.

Kumar, Kavi K.S., Shyamsundar, Priya, and Nambi, Arivudai. 'Economics of Climate Change Adaptation in India.' *Economic and Political Weekly,* 45(18) (2010): 25, 27–29.

Lopez, Ramon. 'The Environment as a Factor of Production: The Effects of Economic Growth and Trade Liberalization.' *Journal of Environmental Economics and Management,* 27(2) (1994): 163–184.

McKinsey & Company. 'Pathways to a Low-carbon Economy.' McKinsey & Company, 2009.

Millin, Mark and Tennassie, Nichola. 'Explaining Economic Growth in South Africa: A Kaldorian Approach.' *International Journal of Technology Management & Sustainable Development*, 4(1) (2005): 47–62.

Nair, Ramachandran P.K., Mohan Kumar, B., and Nair, Vimala D. 'Agroforestry as a Strategy for Carbon Sequestration.' *Journal of Plant Nutrition and Soil Science,* 172(1) (2009): 10–23.Available at: https://onlinelibrary.wiley.com/action/doSearch?ContribAuthorStored=Ramachandran+Nair%2C+P+K

Nelson, C. Gerald, Rosegrant, Mark W., Koo, Jawoo, Robertson, Richard, Sulser, Timothy, Zhu, Tingju, Ringler, Claudia, Msangi, Siwa, Palazzo, Amanda, Batka, Miroslav, Magalhaes, Marilia, Valmonte-Santos, Rowena, Ewing, Mandy, and Lee, David. 'Climate Change: Impact on Agriculture and Costs of Adaptation.' Food Policy Report, International Food Policy Research Institute, Washington, DC, October 2009.

Nordhaus, D. William. 'A Review of the Stern Review on the Economics of Climate Change.' *Journal of Economic Literature,* 45(3) (2007): 686–702.

Nordhaus, D. William, and Yang, Zili. 'A Regional Dynamic General-Equilibrium Model of Alternative Climate-Change Strategies.' *The American Economic Review,* 86(4) (1996): 741–765.

Padukone, Neil. 'Climate Change in India: Forgotten Threats, Forgotten Opportunities.' *Economic and Political Weekly,* 45(22) (2010): 47–54.

Ramsey, F. 'A Mathematical Theory of Saving.' *The Economic Journal,* 38(152) (1928): 543–559.

Richmond, Anthony. 'Reactive Migration: Sociological Perspectives on Refugee Movements.' *Journal of Refugee Studies,* 6(1) (1993): 7–24.

Sanghi, Apurva, and Mendelsohn, Robert. 'The Impacts of Global Warming on Farmers in Brazil and India.' *Global Environmental Change,* 18(4) (2008): 655.665.

Stock, Tim, Obenausb, Michael, Slaymakerc, Amara, and Seligera, Grunther. 'A Model for the Development of Sustainable Innovations for the Early Phase of the Innovation Process.' *Procedia Manufacturing*, 8 (2017): 215–222.

United Nations, Department of Economic and Social Affairs, Population Division. *World Urbanization Prospects: The 2007 Revision Population Database*. New York, NY: United Nations, 2007.

Wells, Heather, and Thirwal, A. P. 'Testing Kaldor's Growth Law across the Countries of Africa.' *African Development Review,* 15(2–3) (2004): 89–105.

Appendix

Table 5.1 Relevant results of the regression analyses with emissions as the dependent variable

	Regression results
Wald $\chi^2(4)$	4.92
Significance level	0.296
Export (Exp) coefficient	0.0798554
z-statistic	0.84
Significance level	0.402
Economic growth (EcG) coefficient	0.0222382
z-statistic	1.24
Significance level	0.216
External debt stock (Dbt) coefficient	–0.1657891
z-statistic	–1.57
Significance level	0.116

Table 5.2 Relevant results of the regression analyses of emission and economic growth with migration

	Regression with pollution (Em) as dependent variable	*Regression with economic growth (EcG) as dependent variable*
Wald $\chi^2(1)$	0.14 0.73	
Significance level	0.7043 0.392	
Migration (Mig)	$4.80e^{-08}$ $5.36e^{-07}$	
coefficient	0.38 0.86	
z-statistic	0.704 0.392	
Significance level		

Table 5.3 Data set of the values of variables under study

Countries	*Years*	CO_2 *emission (metric tons per capita) (Em)*	*Export (% of gross domestic product) (Exp)*	*Economic growth (annual %) (EcG)*	*Ext. debt (% of gross national income) (Dbt)*
Bangladesh	1998	0.190182468	11.75733123	5.177026873	30.1656125
Bangladesh	1999	0.19558517	11.75864239	4.670156368	31.13473307
Bangladesh	2000	0.211802225	12.34420123	5.293294718	28.31641144
Bangladesh	2001	0.242020016	13.38655921	5.077287776	26.92982364
Bangladesh	2002	0.246756218	12.40996893	3.83312394	29.32794849
Bangladesh	2003	0.25660216	11.4311487	4.739567399	29.46910141
Bangladesh	2004	0.266823041	11.14650895	5.23953291	29.02710279
Bangladesh	2005	0.275246594	14.3928419	6.535944941	25.51766945
Bangladesh	2006	0.299529173	16.35346357	6.671868265	26.57517045
Bangladesh	2007	0.301630726	16.99533485	7.058636206	25.46627278
Bangladesh	2008	0.332727819	17.65885848	6.013789759	23.7197055
Bangladesh	2009	0.357158767	16.94013291	5.045124794	22.99902356
Bangladesh	2010	0.393936699	16.02411269	5.571802274	21.57065817
Bangladesh	2011	0.412011186	19.92207496	6.46438388	19.64852433
Bangladesh	2012	0.433487963	20.16158886	6.521435078	19.74463623
Bangladesh	2013	0.442400828	19.53787411	6.013596067	20.9773941
Bangladesh	2014	0.459141965	18.98966	6.061093054	19.33892152
Bhutan	1998	0.703429831	33.10001129	5.914030835	53.3421717
Bhutan	1999	0.690592474	31.11449204	7.983972022	55.27143776
Bhutan	2000	0.690660881	29.37722626	6.93302424	48.23184514
Bhutan	2001	0.653044437	26.2861735	8.203773861	57.27798994
Bhutan	2002	0.689377786	23.87154995	10.72784061	72.97631589
Bhutan	2003	0.605839592	25.52907435	7.664334042	81.15749926
Bhutan	2004	0.481081773	31.02749389	5.896408192	87.56547987
Bhutan	2005	0.603125919	38.24974176	7.12255965	81.33305635
Bhutan	2006	0.583684405	54.41944342	6.849365932	80.64565691
Bhutan	2007	0.571168834	54.96856072	17.92582494	67.89385663
Bhutan	2008	0.601619231	48.79026012	4.768354226	56.60944259
Bhutan	2009	0.544051575	44.70198135	6.657223796	64.78813815
Bhutan	2010	0.67026322	42.45279115	11.73085436	62.41303919
Bhutan	2011	0.990398509	41.20541495	7.890913893	62.8025732
Bhutan	2012	1.086025018	38.72533426	5.071709815	86.20742508
Bhutan	2013	1.203220818	40.4599045	2.142496554	95.71631473
Bhutan	2014	1.289321371	36.28463735	5.745455168	100.6725158
India	1998	0.921501664	11.13422522	6.184415821	23.85741345
India	1999	0.962521771	11.57237585	8.845755561	22.21572554
India	2000	0.979870442	13.13378056	3.840991157	22.12115802
India	2001	0.97169808	12.69031341	4.823966264	20.95807363
India	2002	0.967381128	14.41424036	3.803975321	20.9546578
India	2003	0.992391683	15.10495122	7.860381475	19.97770427
India	2004	1.025027608	18.04674645	7.922943418	17.79802293
India	2005	1.068563218	19.82436588	9.284824616	15.09279297
India	2006	1.121981504	21.66419921	9.263964759	17.47321246
India	2007	1.19320986	21.00759325	9.801360337	17.06141387
India	2008	1.310097833	24.26743306	3.890957062	19.25010182

Countries	*Years*	*CO_2 emission (metric tons per capita) (Em)*	*Export (% of gross domestic product) (Exp)*	*Economic growth (annual %) (EcG)*	*Ext. debt (% of gross national income) (Dbt)*
India	2009	1.431844254	20.61556009	8.479783897	19.47762813
India	2010	1.397008906	22.5904459	10.25996306	17.72344109
India	2011	1.47668635	24.54041132	6.6383638	18.50558011
India	2012	1.598098637	24.53443066	5.456387552	21.73790188
India	2013	1.591437853	25.4308613	6.386106401	23.30366621
India	2014	1.730000432	22.96796301	7.410227605	22.74930817
Maldives	1998	1.116966189	–	7.494683025	37.81721096
Maldives	1999	1.536355489	–	6.175655828	39.27570906
Maldives	2000	1.608654559	–	3.845810392	34.15928575
Maldives	2001	1.609750999	–	–3.943634354	27.55210047
Maldives	2002	2.01825094	–	7.268386313	30.8617452
Maldives	2003	1.674490171	–	13.75004982	29.06000442
Maldives	2004	2.149950229	–	6.033754049	29.07389939
Maldives	2005	1.886198547	–	–13.12905343	33.27549243
Maldives	2006	2.318681252	49.33400091	26.1114935	38.61260279
Maldives	2007	2.324131877	96.54770455	7.713867271	52.672391
Maldives	2008	2.444284083	86.71027947	9.485332661	62.76800624
Maldives	2009	2.492932319	73.0056489	–7.228841465	73.04716962
Maldives	2010	2.56531353	77.56179345	7.265129068	45.55343919
Maldives	2011	2.629542746	88.02790499	8.566733531	40.92228248
Maldives	2012	2.876986973	85.75581078	2.517383942	39.56320556
Maldives	2013	2.749809385	88.39479084	7.281073979	36.97653438
Maldives	2014	3.26955985	89.29190661	7.329606386	38.70868043
Nepal	1998	0.098508398	22.82205122	3.016389482	54.90692529
Nepal	1999	0.138145835	22.84847209	4.412573271	60.06892224
Nepal	2000	0.129282276	23.28400371	6.199999988	52.1902679
Nepal	2001	0.135225857	22.5607505	4.799762644	45.49492791
Nepal	2002	0.106876799	17.73712952	0.120266897	49.66764394
Nepal	2003	0.113901965	15.69986449	3.945037767	50.88020148
Nepal	2004	0.105477444	16.68269526	4.682603245	46.466896
Nepal	2005	0.120277398	14.58368679	3.479181046	39.14125581
Nepal	2006	0.098811563	13.44659096	3.364614788	37.3613966
Nepal	2007	0.099736268	12.85566488	3.411560276	34.65635976
Nepal	2008	0.129223796	12.77582026	6.104639142	29.1894338
Nepal	2009	0.162087331	12.41935419	4.53307872	29.04739308
Nepal	2010	0.187128275	9.582536172	4.81641465	23.49609491
Nepal	2011	0.202491061	8.904030421	3.421828241	20.12962489
Nepal	2012	0.211798007	10.07389298	4.781192258	22.21339069
Nepal	2013	0.237169787	10.68907517	4.128877676	20.66602199
Nepal	2014	0.283538526	11.50508764	5.988984661	20.48550064
Pakistan	1998	0.73845596	16.48479118	2.550234295	52.38565507
Pakistan	1999	0.741358828	15.35349933	3.660132744	54.69303838
Pakistan	2000	0.768458119	13.44132462	4.260088012	45.2192214
Pakistan	2001	0.764701583	14.65953961	1.982484032	44.7148306
Pakistan	2002	0.788667608	15.2236172	3.224429973	46.50767147

(*Continued*)

Table 5.3 (Continued)

Countries	*Years*	*CO_2 emission (metric tons per capita) (Em)*	*Export (% of gross domestic product) (Exp)*	*Economic growth (annual %) (EcG)*	*Ext. debt (% of gross national income) (Dbt)*
Pakistan	2003	0.804958723	16.71896741	4.846320935	42.48099647
Pakistan	2004	0.872801659	15.6668995	7.368571359	36.28085315
Pakistan	2005	0.887768063	15.68949543	7.667304271	30.43633592
Pakistan	2006	0.929856974	14.13396282	6.177542036	26.59714137
Pakistan	2007	0.991029936	13.21461341	4.832817277	27.10321835
Pakistan	2008	0.972050346	12.38231088	1.701405465	28.55732408
Pakistan	2009	0.950832112	12.39575242	2.831658519	32.64860364
Pakistan	2010	0.946268168	13.51626785	1.606691959	34.10065704
Pakistan	2011	0.929801105	13.96666966	2.74840255	28.88741717
Pakistan	2012	0.918978434	12.39666276	3.50703342	26.30198459
Pakistan	2013	0.904315983	13.27715343	4.396456633	23.79832281
Pakistan	2014	0.896264105	12.24258363	4.674707981	23.67940024

Source: World Bank database

Table 5.4 Data set of the values of variables under study for regression analysis between pollution (represented by CO_2 emissions) and migration

Countries	*Years*	*CO_2 emissions (metric tons per capita) (Em)*	*Migration (number of people) (Mig)*
Afghanistan	1997	0.059648382	–379474
Afghanistan	2002	0.048715548	929118
Afghanistan	2007	0.085417506	–777497
Afghanistan	2012	0.350370581	448007
Bangladesh	1997	0.202365814	–756121
Bangladesh	2002	0.246756218	–1541457
Bangladesh	2007	0.301630726	–3570954
Bangladesh	2012	0.433487963	–2526483
Bhutan	1997	0.742063417	143
Bhutan	2002	0.689377786	29002
Bhutan	2007	0.571168834	16829
Bhutan	2012	1.086025018	10000
India	1997	0.92007238	–716900
India	2002	0.967381128	–1950910
India	2007	1.19320986	–2913829
India	2012	1.598098637	–2578213
Maldives	1997	1.24881279	–1100
Maldives	2002	2.01825094	13811
Maldives	2007	2.324131877	17994
Maldives	2012	2.876986973	21916
Nepal	1997	0.124308954	–464703
Nepal	2002	0.106876799	–797844
Nepal	2007	0.099736268	–1023896
Nepal	2012	0.211798007	–372369

Countries	*Years*	CO_2 *emissions (metric tons per capita) (Em)*	*Migration (number of people) (Mig)*
Pakistan	1997	0.73370119	–728632
Pakistan	2002	0.788667608	–671488
Pakistan	2007	0.991029936	–1396377
Pakistan	2012	0.918978434	–1181920

Source: World Bank database

Table 5.5 Data set of the values of variables under study for regression analysis between economic growth and migration

Countries	*Years*	*Eco growth (annual %) (EcG)*	*Migration (number of people) (Mig)*
Afghanistan	1997	–	–379474
Afghanistan	2002	–	929118
Afghanistan	2007	13.74020499	–777497
Afghanistan	2012	14.43474129	448007
Afghanistan	2017	2.595542453	–299999
Bangladesh	1997	4.489896497	–756121
Bangladesh	2002	3.83312394	–1541457
Bangladesh	2007	7.058636206	–3570954
Bangladesh	2012	6.521435078	–2526483
Bangladesh	2017	7.284208377	–2350001
Bhutan	1997	5.373838519	143
Bhutan	2002	10.72784061	29002
Bhutan	2007	17.92582494	16829
Bhutan	2012	5.071709815	10000
Bhutan	2017	6.818777705	0
India	1997	4.049820849	–716900
India	2002	3.803975321	–1950910
India	2007	9.801360337	–2913829
India	2012	5.456387552	–2578213
India	2017	6.624227116	–2449998
Maldives	1997	8.39794083	–1100
Maldives	2002	7.268386313	13811
Maldives	2007	7.713867271	17994
Maldives	2012	2.517383942	21916
Maldives	2017	8.828456057	9758
Nepal	1997	5.048612536	–464703
Nepal	2002	0.120266897	–797844
Nepal	2007	3.411560276	–1023896
Nepal	2012	4.781192258	–372369
Nepal	2017	7.499402289	–350000
Pakistan	1997	1.014396014	–728632
Pakistan	2002	3.224429973	–671488
Pakistan	2007	4.832817277	–1396377
Pakistan	2012	3.50703342	–1181920
Pakistan	2017	5.700621241	–1071778

Source: World Bank Database

6 Climatic refugees

Internal migration in the face of climate change in India

Srijan Banerjee[1]

In the years following World War II, the global economy has experienced radical changes that have resulted in development and amelioration of the well-being of a considerable population. A number of new nations have emerged as a result of attainment of independence from colonizers, thus acting as a step towards liberation from conditions of abject living standards. However, in the milieu of this development and upliftment narrative, a persistent phenomenon has been tightening its grip in the background. In the pursuit of economic growth and development, very little attention had been devoted to the gradual change in climatic conditions occurring incessantly across the globe. In fact, many economists have tried to establish causality between the economic progressions of a nation and the deteriorating condition of its environment. However, this fact has been continually pushed into denial, and as a result, the ubiquitous practice of environmental degradation persists without any disruption. This, in turn, escalates into a situation where a serious natural disaster occurs, causing mass jeopardy. Generally speaking, a number of anthropogenic factors are held responsible for bolstering this temporal variability of climatic conditions, which in turn begets far-reaching consequences. In contemporary times, the change in global climatic conditions and the resultant ramifications are considered an issue receiving unequivocal consensus and demanding systematic interventions across international borders. Amidst the consensus about the occurrence of climate change, debates continue on the issue of its extent and repercussions. According to the fifth assessment report of the Intergovernmental Panel on Climate Change (IPCC), paradigm shifts in the climatic profiles of many regions across the globe are a cause of disquieting apprehension and need immediate attention (IPCC, 2014b). According to the Stern Review (2007), climate change is a phenomenon that will continue to persist for the next few decades and, as a result, there is supreme urgency to take necessary action to build resilience and diminish the costs (see also Foresight, 2011). The colossal economic costs associated with climate change (almost 5% of the global GDP per year) and the significant quantum of population likely to be displaced, both temporarily and permanently, are issues that have raised concerns. The growing incidence of changing climatic conditions upon a large quantum of the world population has raised concern about millions who are being forced to leave their homes and migrate elsewhere. This chapter primarily focuses on the issue of region-wise vulnerability of India to persistent climatic events, including both slow- and rapid-onset events and their resultant implications on the dislocation of a large portion of the population, escalating internal migration.

There have been myriad attempts to discern the relationship between climate change and human migration, as is evident from the vast literature. One thing that can be observed from the discourse is that human mobility has been viewed as an obvious coping mechanism in the face of the adversities that changing climatic conditions have to offer. However, debates crop up while trying to quantify the moving population or while trying to theorize it in light of conflicting empirical results. The United Nations Refugee Agency United Nations High Commission for Refugees (UNHCR) has indicated that displacement due to climate change is a contemporary reality and about 21.5 million people have been displaced annually due to some form of weather-oriented sudden-onset events since 2008 (UNHCR, 2016). A comprehensive estimate by Myers (1996) suggests that about 25 million environmental migrants were present in 1995, and the number may well escalate to 200 million by 2050 (Myers, 2002). Although contentions still exist regarding the accuracy of such predictions, nevertheless they provide us with the possible extent and enormity of the problem. Also, one should also be aware that climate change may influence human movement through rapid-onset events like floods, storms and earthquakes or slow-onset events like desertification, changes in precipitation and rise in sea levels (Beine and Parsons, 2015). Taking into account climatic events like earthquakes, hurricanes, floods and droughts, Ruiz (2017) clearly indicated that both drought and abundant precipitation are significant push factors for internal migration in Mexico and manifest autonomously of socio-economic and political factors.

On the other side of the literary spectrum, many academicians strongly oppose the reductionist approach of relating climate change with human migration. According to Piguet et al. (2011), environmental factors cannot be singled out as drivers of human migration, since they remain in complex interplay with other decisive factors of a socio-economic, political or demographic nature. A number of papers have actually tried to look at migration as a result of climate change through the channel of perturbations in the agricultural setup of a region. Falco et al. (2018) support this link by postulating that climatic shocks have a significant impact on agriculture and agriculture is the mainstay of people living in rural areas of developing countries, the major source of migrants worldwide. Despite being considered a global concern, Black et al. (2011) have pointed out that, although the incidence of environmental changes on human migration will increase with time, migration decisions can be viewed as an effective method of adaptation and establishing long-term resilience. They have also concluded that apart from the incidence of changing climatic conditions across geographical spaces, prevalent socio-economic conditions also play a pivotal role in determining the extent of vulnerability of a resident population to climate change. This result is particularly intriguing, since it has a contextual connotation to it and clearly showcases the route through which different economic and social factors are responsible for different levels of vulnerability worldwide. This can be readily understood by observing the findings of Munshi (2003), where, in an attempt to investigate network effects amidst Mexican migrants in the US labour force, the paper explored a strong negative correlation between rainfall at origin and the decision to emigrate to the United States (see also Feng et al., 2010).

A fact that has surfaced prominently in recent literature is that developing countries are particularly susceptible to climate change. According to Panda (2010), weak ecological environment, high vulnerability of economic systems to risk and restrictions of a lower-income population in pursuing suitable adaptation mechanisms are factors that can be held responsible for this susceptibility. In terms of susceptibility, India is a country which has drawn considerable global attention. Many studies have concluded unequivocally that India would eventually emerge as one of the most direly affected countries in the face of climate change. The climatic disparities experienced by different parts of the country have espoused different types and manifestations of changes in climatic conditions across its geographical extent. Rising temperatures and changing trends of precipitation are also being observed in many areas of the country which threaten to affect vulnerable communities with losses of occupation and income and food insecurity and eventually result in mass displacement. In this scenario, the problem of climatic migrants is already in existence and is expected to worsen many times over in the years to come. However, this issue seems to have drawn very little attention both from the government as well as academicians. The problem of climatic refugees is a poorly documented issue in India, with very few studies and papers delving into the matter. According to the Internal Displacement Monitoring Centre (IDMC), in India, about 1,346,000 fresh cases of displacement occurred in 2017, a fact that is congruent with earlier estimates (IDMC, 2017). However, no major effort has been made by government agencies to formulate a plan of action to counter this problem and demarcate the areas which account for the majority of the migrants. Situated against such a background, the major objective of this chapter is to provide a descriptive account of the issue of changing climatic scenarios and their impact on the dislocation of a large quantum of population in India. In doing so, the chapter has delved, to a great extent, into cases of both slow- and rapid-onset climatic events and their respective impacts on human displacement. The chapter also aims to provide a comprehensive analysis of anthropogenic factors such as unplanned urbanization and continual human settlement in vulnerable areas in contributing to further exacerbation and increased susceptibility.

Analytical methodology

In order to deliver a comprehensive analysis, cases of both slow-onset and rapid-onset climatic events are taken into account for evaluation and assessment. For the slow-onset event, the case of the changing morphology of the river Ganga and its impact on human displacement in the Malda district of West Bengal, India, is considered. This issue is of particular significance, as, although the change in morphology of the river Ganga is a gradual process that has been ongoing for a long period of time, it stands as a major cause of regular rapid-onset natural disasters like soil erosion, bank failure and seasonal flooding. As a result, a large portion of the district's population is vulnerable to loss of shelter and occupation, particularly those who reside at or near the left bank of the river Ganga. For the analysis, the causes and extent of the shift in stream position of the river Ganga are provided, along with an account of the resultant displacement in two of the most vulnerable blocks: Manikchak and Kaliachak II. A mouza-level analysis is provided as well, in order to identify the affected mouzas and the extent of the damages across time for both the Manikchak and Kaliachak II blocks using data from Guchhait (2018). Also, estimates of the displaced population as well as affected Gram Panchayats in the respective blocks are presented using data from the disaster management plan of Malda district (2017). For the rapid-onset event, the case of the Tamil Nadu floods of 2015 and the resultant population displacement is considered. This issue is very pertinent in contemporary times, since cases of urban flooding have been on the rise over the past few years in India. In this analysis, a depiction of the precipitation anomaly is provided using secondary data from the Indian Meteorological Department, along with an account of the resultant deluge and the affected districts. In order to exhibit the extent of economic and social damages, data from the Joint Detailed Needs Assessment Report of Sphere India (2016) is used, with a district-level analysis. Finally, the trend of rapid and unplanned urbanization is highlighted as a major cause of exacerbation of the flood situations, using data from various Census years, and it is observed that the refuge-seeking population is mostly composed of economic migrants who had settled earlier in urban vicinities in search of better occupations and basic amenities.

Climatic refugees: the debate

Although categorized as a contemporary and newly articulated issue, the concept of migration as a result of environmental changes found a place in 19th-century literature and remains an extant phenomenon. Ravenstein (1889) accounted for a stream of continuous migration as a result of 'unattractive climate' and onset of changes in climatic conditions. However, from then to the early 1980s, the bulk of the literature on migration has failed to incorporate or even hint at environmental and climatic factors and inter-temporal changes in them as plausible explanatory routes. Core theories of human migration like Lewis (1954), Harris-Todaro (1970), Todaro (1976) and Zelinsky (1971) have all proposed explanations with economic factors as the main area of concern.

The concept of migration as a reaction to environmental change resurfaced in the literature with the coinage and subsequent popularity of the term 'environmental refugees' by Lester Brown of the World Watch Institute in the 1970s. El-Hinnawi (1985) professed a comprehensive and arguably the most popular definition of 'environmental refugees' in his reports of the United Nations Environmental Programme (UNEP), where he defined environmental refugees as the quantum population who are forced to abandon their traditional settlements, temporarily or permanently, due to some environmental disruption that proves a jeopardy to their existence or standard of living. Bates (2002), on the other hand, points out that previous academic efforts to provide a satisfactory definition of climatic refugees have been vague. He defines environmental refugees as people compelled to migrate as a response to change in 'their ambient non-human environment' and goes on to categorize the movement based on three causes: disaster, expropriation and deterioration. Myers (2002) described the term 'environmental refugees' as a new phenomenon in the global scenario and propounded that they are unable to enjoy a secure livelihood in their place of dwelling due to

> drought, soil erosion, desertification, deforestation and other environmental problems, together with the associated problems of population pressures and profound poverty.

According to his estimates, in 1995, the total number of environmentally displaced people was close to 25 million worldwide, as opposed to approximately 27 million refugees who were absconding 'political oppression, religious persecution or ethnic troubles'. He further predicted that the number of environmental refugees could swell to 50 million by 2010 and further burgeon to about 150 million by the end of the 21st century. (Myers, 1993; see also Myers, 1997; Myers and Kent, 1995; Blaikie et al., 1994). The narrative of the argument remained an alarmist prediction based on conjectures rather than being empirical regarding contemporary human movement (Piguet et al., 2011). However, on the other side of the debate were academicians who had strong objections towards the term 'environmental refugees'. Homer-Dixon (1999) strongly critiques the loose and insincere manner in which terms like 'environmental refugees' or 'environmental migrants' are used in academia, particularly in studies depicting human movement, making their definition hard to comprehend and interpret. He even harps on the argument that migration only due to environmental tensions occurs rarely, since environmental factors manifest through complicated interactions with other factors and the migration decision cannot be linearly represented as a function of some pull and push factors only. Many academics like Black (2001), Kibreab (1997) and Castles (2002) have also had disputes over the accuracy and causality of environmental implications over migration. They profess the notion that human migration is a social and economic as well as ecological phenomenon and cannot be reduced to only one causal determinant (Oliver-Smith, 2009). According to Black (2001), environmental migration is nothing but a 'cyclical coping' mechanism practiced by the global population for centuries and is a constituent of overall adaptation rather than only responding to changes in the environment.

From the legal standpoint, as well, the concept of 'environmental refugees' remained a contentious topic of debate. The United Nations Convention pertaining to the status of refugees (1951) legally defined a 'refugee' as a person who leaves his/her nation for fear of persecution based on race, religion, membership of an ethnic or social group or political opinion. Therefore, human displacement for environmental reasons is not accounted for in this definition of refugees. As a result, Hugo (1996) proposed the usage of the term 'environmental migrant' and acknowledged the role of environmental change in bolstering involuntary migration. It was also feared that the term 'environmental refugee' would eventually masquerade as political displacement and would give the state an alibi not to grant asylum. Kibreab (1997) claimed that the term was created in order to 'depoliticize the causes of displacement'. Therefore, although there is general scientific consensus over the phenomenon of change in climatic conditions and the hazards entailed by it, academicians continue to contest the definition and implications of the quantum of global population displaced due to climatic reasons.

Slow-onset climatic drivers of migration: change in morphology of the river Ganga and the resultant displacement in Malda, West Bengal, India

Before progressing on to broader details, one must be aware of the concept of slow-onset climate change. Slow-onset climate change is a global phenomenon characterized by environmental events along with changes in climatic conditions whose impacts are not felt immediately. The process of occurrence of such events is gradual, spread across a long span of time. In this analysis, the gradual change in the morphology of the river Ganga has been highlighted, along with shifting of its course and the resultant human displacement in the Malda district of West Bengal, India. Although, categorized as a slow-onset natural event, this change in morphology is the root cause of regular rapid-onset natural disasters like floods, bank failure, soil erosion and other events that leave a large section of the district's population susceptible to loss of shelter and livelihood.

District profile at a glance

Sharing a 165.5-km-long international border with Bangladesh, the Malda district of West Bengal, India (see Table 6.1 in the appendix for some detailed information about its administrative division) is considered unique due to its location and as a seat of major international trade. Spread over an area of 3733 sq km between 24°40'20"N to 25°32' 08"N latitude and 87°45'50"E to 88°28'10" E longitude, the district is home to a total population of 3,988,845, which is about 4.41% of the total state population of West Bengal (Census, 2011a).

Soil erosion and flood situations in Malda

One of the most persistent phenomena observable in Malda district is river bank erosion and resultant floods. This process of fluvial dynamics sometimes alters the morphology of the stream over time, and ultimately the stream loses its profluence, forming a complex network of channels and various associated habitats (Kondolf, 1997). Bank failure is also predominant in this area, defined as the breakdown of the banks of a stream or a river due to instability in their slope or when the latent pressure exerted on a bank decimates the cohesiveness of the constituent soil particles. Apart from natural causes, anthropogenic factors like erection of dams, barrages and bridges and construction of embankments and bulwarks play a huge role in speeding up and inflating river bank erosion, as they interferes with the natural dynamics of the river course. The different erosion classes based on intensity and area covered (see Table 6.2 in the appendix) show that about 21.69% of the total land area of Malda is vulnerable to moderate and in some cases severe soil erosion.

As one would expect, incessant erosion leads to vulnerability of a large quantum of the district's population, especially in the floodplain regions of the rivers, both with respect to security of dwelling units and safety of livelihood and basic amenities (DHDR, Malda, 2007). Since the floodplain area is a lucrative space for settlement owing to its high soil fertility and other facilities, human settlement is widespread. This entails a negative externality, since increasing human settlement, along with establishment of colonies, transportation and agricultural activities, creates pressure on the local soil resources, and as a result, the soil starts to lose its cohesive character and ultimately falls prey to riverine dynamics. Each year, the landless population seeks government reallocation where the claims often exceed the available land resources of the government. As a result, many of the affected are forced to take refuge in adjoining areas where the original inhabitants are often not ready to welcome them. A state of tension thus ensues between the affected and non-affected people, and the victims are left with no choice but to settle in temporary huts along highways and roads (Thakur et al., 2011). It must also be kept in mind that due to the perpetual nature of the problem, the impacts suffered by the affected population remain indelible, and it is a difficult task to replenish the losses and provide sufficient reallocation. The problem of migration and displacement entails an economic facet, because the decision to migrate has causality with economic affluence. In other words, an individual faced with abject poverty may not have the resources to actually progress to moving out and as a result is forced to stay in a space that is riddled with environmental disasters. Also, in the case of Malda, the decision to migrate may not always be taken due to loss of habitat only but also depends on the extent of loss of public goods and institutions like roads, schools, banks, post offices and so on, along with security of livelihood. Therefore, climatic factors manifest both directly and indirectly (through economic channels) in the region.

In Malda, the major erosion-prone areas appear to be consistent; that is, they are all located on the left bank of the Ganga and spread across the Diara blocks (the most fertile area with accretions of Gangetic alluvium) of Manikchak and Kaliachak II and III blocks before advancing on to parts of Murshidabad. As the most juvenile and fertile plain of Malda, the soil profile of the region is a concoction of thin accretions of silt over thick layers of loosely packed loamy soil and thus is vulnerable to undermining by the river flow and tends to break away in large portions.

Change in morphology of the Ganga river

The temporal change in morphology of the river Ganga is a major issue of concern for the inhabitants of the district, especially in the left bank areas. The accumulation of millions of tons of sediment load along with river bed deposits engenders problems such as stymied flow and dwindling of the depth of the river bed, thus having a percolating effect on direction and pace of flow. As a result of such sedimentation and changes in flow character, coupled with inconsistent discharge, the river casts lateral pressure on the walls of the banks and invites erosion in Malda (Mondal et al., 2016; Thakur et al., 2011). The morphology of the river Ganga has changed significantly over time and can be deduced from subsequent reconstructions of stream positions, where there is an indication of a continuous eastward shift of the channel (Figure 6.1) (Mandal, 2017; Majumdar and Mandal, 2018). There have also been high levels of sedimentation, since a sizeable amount of land accretion is observable across time. A noticeable feature is that two additional channels have emerged from the main stream, and the area of the intermediate island has consistently increased due to high levels of sediment accretion (Laha and Bandyapadhyay, 2013). It must also be noted that the stream is steady at two points, Rajmahal Hills, as it acts as a steady geological obstacle, and Farakka Barrage, as an anthropogenic obstruction. The major shift in the stream has occurred between these two stable points (Sinha and Ghosh, 2011; Mondal et al., 2016). This change is primarily attributed to incessant erosion of banks and has been persistent

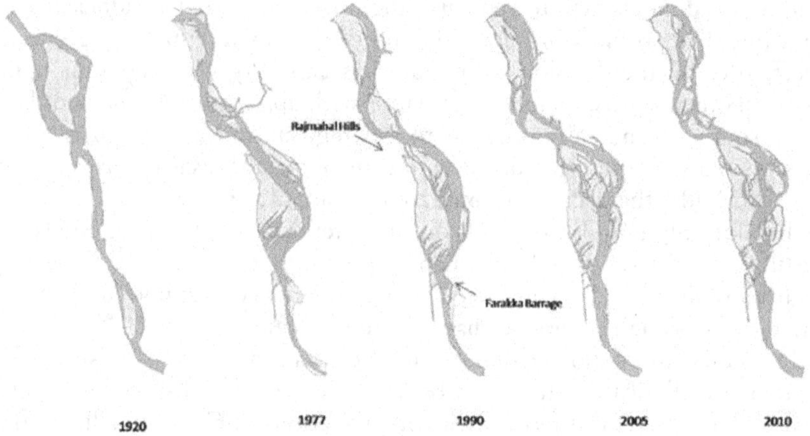

Figure 6.1 Temporal changes in the course of the river Ganga
Source: Laha and Bandyapadhyay (2013)

across a long period of time. Therefore, a short-run scrutiny of the situation will not be enough to perceive the actual scenario. For this reason, this issue has not received adequate administrative and bureaucratic attention, except in the case of resultant rapid-onset disasters, when the need for emergency disaster management has emerged.

Having said this, one must also note that the prevalent erosion on the left bank of the Ganga is not purely due to environmental reasons like the shifting nature of the river stream but also has crucial anthropogenic reasons behind it (Rudra, 1996a, 1996b) (see also Majumdar and Mandal, 2018). The erection of the Farakka Barrage in 1971 has had a consequential impact in escalating the problem of river-based erosion, especially on the left bank of the river Ganga. The 2.64-km-long barrage was devised to channelize 40,000 cusecs of river water in the direction of the Bhagirathi River so that the sediment burden could be flushed into the deeper part of the estuary and as a result ameliorate the navigability of the Kolkata port (Rudra, 2004a). But in reality, Farakka Barrage acts as a fetter to the normal trajectory of the river, thus compelling it to make its own way, resulting in the shifting of its natural ambit and rampant erosion in the upstream region of the barrage. This phenomenon has been a persistent problem in the region for the last four decades or so, as is evident from the detailed analyses of Rudra (1996a, 1996b, 2004a, 2004b, 2010).

In this analysis, the cardinal focus is laid upon two community blocks, Manikchak and Kaliachak II, which have been the worst-hit regions due to the aforementioned phenomenon. In this region, bank failure is a biannual phenomenon: pre-flood failure occurring due to increasing lateral pressure on banks and post-flood failure when stagnant flood water penetrates the soil, making it susceptible to aggregative collapse when the water recedes (Keshkar et al., 1996; Thakur et al., 2011). It has been estimated that about a 750-sq km area has been wiped off the map in the last 30 years due to erosion. The point of analysis lies in the constant displacement of resident families and the decimation of vital institutions and public goods due to loss of land cover and floods.

(i) Manikchak block

Located in the Sadar Division of Malda, Manikchak block spreads across an area of 316.39 sq km with a total population of 269,813 (Census, 2011b). It is located in the Diara region of the district and is the worst-affected block with regard to erosion and floods. A brief block profile is presented in Table 6.3 in the appendix. Although located in a region abundant in fertile alluvial soil, the agricultural advancement in the region appears to be stunted due to the constant flood and erosion activities of the Ganga. In fact, frequent water logging, crop losses, loss of livestock and considerable depreciation of cultivable land are widespread in Manikchak. Therefore, the fate of a sizeable quantum of the residing population remains dependent on fluvial dynamics. According to the District Human Development Report of Malda (2007), the total area eroded until 2006 was 13,204.02 acres, with 3330 families losing their settlements. Along with this, the inhabitants of 236 riverside villages had lost their share of cultivable lands, and an approximate estimate of 5043 ha of prime cropland was decimated in the Diara region. Therefore, one can imagine the economic conditions of the displaced families, who were flung into a state of economic destitution overnight. Additionally, due to a high level of settlement density in the region, access to fresh cultivable land is becoming increasingly difficult, which in turn confines the scope and extent of occupational choice for the displaced population. The mouzas affected by Gangetic erosion, along with the extent of the damage across three and a half decades (Table 6.4 in the appendix), show that the total number of affected mouzas has remained more or less consistent, with mouzas like Sobhanathpur, Mirpur, Bagdukra and Dergama experiencing a land loss of over 60% land area for more than a decade. The estimates corresponding to 2017 depict an alarming situation where 7 out of the total 23 affected mouzas had lost more than 60% of their land area owing to encroachment and erosion. Besides, across time, the number of mouzas experiencing land loss in the range of 30% to 60% has consistently remained quite large, thus reflecting the grave situation of the overall block. A striking feature of the river Ganga that can be observed from this mouza-level analysis is that there is irregularity in the engulfing of the land area. Put another way, due to the oscillating nature of the river within the jurisdiction of the meander belt, one gets different estimates for land loss (Mandal 2017). In terms of land area engulfed by the Ganga, an escalating tendency can be observed across time in Manikchak block.

In terms of gauging the population affected due to the persistent change in the morphology of the Ganga and resultant disasters, one gets a clear indication of the quantum displacement. As can be observed from Table 6.5 in the appendix, Dakshin Chandipur Gram Panchayat (GP) has recorded the highest number of affected people, although it falls under the category of 30% to 60% land area engulfed by the river Ganga. This fact suggests that in Dakshin Chandipur GP, the majority of the land that has been lost to the persistently encroaching Ganga river were the settlements of a sizeable proportion of the total populace of Manikchak. Also, the location of the block is such that it is a major point of eastward shift of the Ganga river which has been occurring for more than three and a half decades. GPs like Manikchak, Hiranandapur and Dharampur were already in the more than 60% land loss area category and have recorded a large number of affected inhabitants as a result. According to the disaster management plan of Malda district (2017), an additional 40,000 inhabitants of Manikchak block have been identified as under risk and vulnerability, along with an important public health centre at Bhutni. Another important feature prevalent in Manikchak block is that at times of land erosion, the displaced population is forced to move inwards into adjoining areas, thus having a major impact on the local resources of that area. As a result of the migrant population, the land resources and land use pattern of a particular area experience changes. Therefore, the problem of human displacement and its implications need to be observed in totality rather than in isolation during emergency situations demanding disaster management.

(ii) Kaliachak II block

Located in the Diara region of Malda, Kaliachak II block spreads across 209.17 sq km with a total population of 210,105 according to Census 2011 estimates. The majority of the block consists of rural areas where agriculture is the main source of occupation. A block profile is presented in Table 6.6 in the appendix.

Like Manikchak, Kaliachak II is also centred on the agricultural sector, which has faced considerable jeopardy as a result of seasonal floods and erosion. Located on the left bank of the river Ganga, Kaliachak II has been a major point of transition of stream position for many decades. According to the District Human Development Report of Malda (2007), the total area eroded until 2006 was 25,114.67 acres, with 7378 families losing their settlements and being forced to move elsewhere for shelter. Out of the 43 mouzas, 20 were announced as affected as a result of the ongoing process of erosion and bank failure. Therefore, as one can imagine, the magnitude of damage incurred by the resident population is quite large. The conditions that cause human movement manifest mainly through two major avenues.

First, due to land erosion, a large population loses their settlements and is forced to move to other areas in search of shelter. Second, the persistent loss of agricultural land and crops is considered a major blow to the resident population, and as a result, they have no choice but to search for more lucrative settlements. A basic feature of Kaliachak II block is that it mainly consists of rural areas where the level of development of infrastructure and basic amenities is more or less uniform. Therefore, migrant populations often choose to settle outside the block rather than within it. However, a persistent problem in this block is that in the face of devastating floods and soil erosion, many people cannot move elsewhere and are forced to stay in the susceptible areas. This trapped population continues to remain in underprivileged conditions with a constant risk of falling prey to such disasters. However, a positive feature of this region is that apart from land loss, Kaliachak II is an area which experiences substantial levels of accretion of silt and sediments and formation of mid-channel bars (chars). As a result, a considerable amount of land loss is compensated. However, the problem of loss of agricultural land and crop-yielding spaces still continues to ravage the block.

A decadal analysis of the affected mouzas and the area engulfed by the Ganga river (Table 6.7 in the appendix) suggests that the number has remained quite uniform across time. However, one can observe that the percentage of the area engulfed by the Ganga river has experienced considerable fluctuations across three and a half decades. One of the severely affected mouzas of Kaliachak II block is Dogachhi, which has persistently recorded land area of more than 60% being engulfed by the Ganga. Other mouzas like Islampur and Char Babupur have experienced fluctuating estimates of land loss across time. The majority of the mouzas affected by the Ganga River have a rural setup where agriculture is the chief source of income of the resident population. As a result, frequent episodes of cultivable land loss as well as loss of crops play a significant role in causing deterioration of the prevalent conditions of sustenance in the block.

A Gram Panchayat-level analysis of the affected population (Table 6.8 in the appendix) suggests that the maximum numbers correspond to Panchanandapur I and II, GP, which had recorded a loss of less than 30% of its land resources to the Ganga river in 2017. According to the disaster management plan of Malda district (2017), an additional population of 10,000 is currently living in a situation of vulnerability to being displaced. However, there seems to be another problem regarding the settlement pattern of the displaced population. As mentioned earlier, Kaliachak II is a block that has experienced soil erosion and loss of land cover on the one hand and considerable volume of soil accretion on the other. This continual process has led to the development of channel bars or chars (accreted volumes of sediment and soil) on the opposite bank of the river Ganga. These spaces of accreted sediments tend to become places of settlement for the already displaced population, who are in search of refuge. However, due to the location of the chars near the territorial border of West Bengal and Jharkhand, there remains a dispute over the administrative jurisdiction of these regions and, as a result, a sense of crisis of identity prevails between the inhabitants of these char regions (see Das et al., 2014).

Rapid-onset climatic drivers of migration: Tamil Nadu floods, 2015 and the resultant population displacement

Rapid-onset climate drivers are characterized by environmental events whose manifestations are abrupt and have immediate impacts and implications. These events may be meteorological in nature like cyclones and hurricanes or geophysical like earthquakes, tsunami and volcanic eruptions or can be attributed to hydrological origins like floods. In fact, speaking from an economic point of view, human migration due to rapid-onset climatic events does not depend only on the intensity or extent of the events but also on the interaction of factors like degree of exposure of the local population to such events and their adaptive capacity (UNISDR, 2017).

In this part of the chapter, the main focus is on the rampant floods that the state of Tamil Nadu faced in 2015, which acted as a major rapid-onset stimulus for quantum displacement. This phenomenon, as many academicians argue, can be put forward as a classic example of meteorological hazards coupled with a significant role of anthropogenic factors. The rapid rate of urbanization in the state, repetitive encroachments near water sources, a significant manifestation of the El Niño phenomenon and lack of necessary preparation and precautions of the administration have all been cited as silent culprits to this exigent situation. In fact, the floods of 2015 are popularly termed 'urban floods', since a majority of urban areas of Tamil Nadu, especially Chennai, were inundated, with copious amounts of losses of vital urban infrastructure and services. Urbanization is a major reason for development of catchments, resulting in inflation of the flood volumes by up to six times (NDMA, 2010), with Chennai being no exception. The analysis of the 2015 Tamil Nadu floods and its exacerbating impacts has been documented comprehensively in literature such as Szynkowska (2016) and Narasimham et al. (2016), and these are the major sources of motivation and reference for this analysis.

A brief geographic profile of the region

As the 11th-largest state in India with respect to area (130,060 sq km), Tamil Nadu ranks sixth in the country in terms of population (7.21 crores per Census 2011). Being a coastal state, Tamil Nadu enjoys a tropical maritime climate with very little contrast between summer and winter temperatures. However, the climatic profile of the region varies from being sub-humid to semi-arid, depending upon the proximity to the sea. The state is mainly dependant on the monsoon rains and is therefore susceptible to occasional droughts in case of monsoon failure. It has also been observed that during drought situations, internal migration increases from rural areas to urban areas, mainly due to food shortages and a sudden inception of economic uncertainty. Chennai, the capital of Tamil Nadu, happens to be the fourth most populated metropolitan city in India and is also designated a fully urban district. Speaking from a topographic perspective, Chennai is a low-lying coastal plain with an average elevation of 6.7 metres above sea level and is drained by two vital rivers, the Cooum in the centre and the Adyar in the south. However, both rivers have reached a near-stagnation stage and prove insufficient in supporting the local population except during the monsoons. Also, due to its plain terrain, the city of Chennai has very little scope for natural runoff, which acts as a basis for aggravating the problem of floods and water stagnation.

Precipitation anomaly and the deluge

During the period between November and December 2015, the southeastern part of the Indian mainland, particularly the states of Karnataka, coastal Andhra Pradesh and Tamil Nadu, experienced heavy downpours. This was the result of a strong and persistent depression over the Bay of Bengal, which owed its build-out to a strong El Niño phenomenon. Although the retreating monsoons contribute to about 30% of the total rainfall received by the state, the precipitation recorded during November–December 2015 far exceeded the normal rainfall estimates. This widespread anomaly caused severe flooding and inundations, especially in the coastal districts like Chennai, Kancheepuram, Thiruvallur, Villupuram and Cuddalore, resulting in rampant economic and social damage.

In November 2015, two spells of incessant rainfall ensued in Tamil Nadu, one from 8th to 9th November and the other from 15th to 17th November. Figure 6.2 stands as a clear indicator of the ground-level scenario, with major upward kinks in the daily mean rainfall estimates experienced on the aforementioned dates. During this time and until the end of the month, Chennai district was one of the worst-hit regions, with rainfall of almost 1200 mm, which is almost three times the average November rainfall estimates (407.4 mm). This was the highest recorded rainfall in November for the region in a century. Also, Cuddalore district was severely affected during this time due to its location near several water bodies which overflowed after reaching the maximum capacity. More than 100 water channels breached in the vicinity of Cuddalore, causing extensive damage to agricultural lands, settlements and basic infrastructure (Szynkowska, 2016).

The third spell of rainfall occurred in the first week of December, with grave impacts on Chennai and other coastal districts like Thiruvallur, Cuddalore, Villupuram and Kanchipuram. Chennai city received 290 mm rainfall on 1st December 2015, the highest rainfall in 24 hours since 1901 (261.6 mm). This resulted in a state of catastrophic deluge to follow suit.

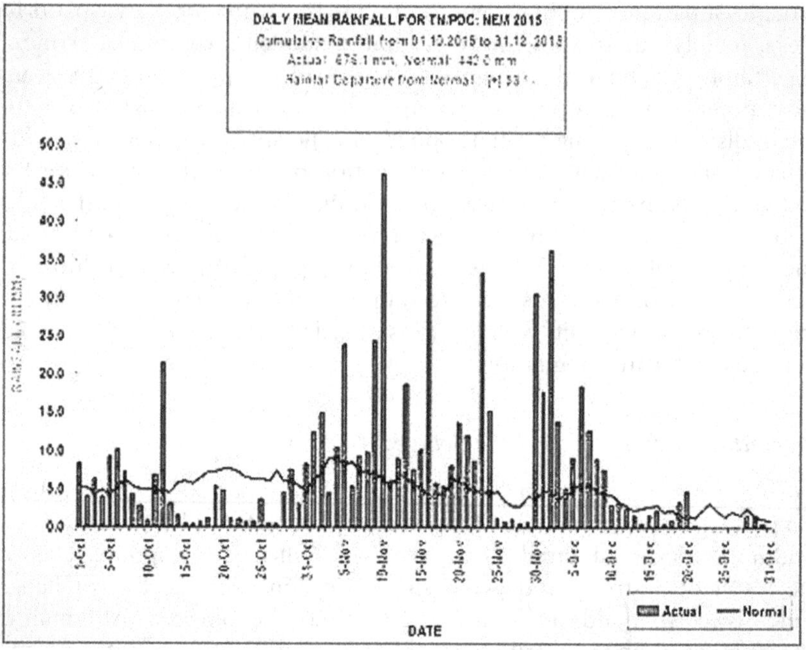

Figure 6.2 Daily mean rainfall in Tamil Nadu and Puducherry from October to December, 2015

Source: Indian Meteorological Department (IMD), Chennai

There was documented evidence that during the last part of the concluding week of November and the first week of December, the depression prevalent over Tamil Nadu and adjoining areas had augmented severely, with Chennai and suburbs crossing the 500-mm mark for rainfall observed. Additionally, discharge of water from the Chembarambakkam reservoir by the district administration exacerbated the situation to a large extent. The reservoir had reached its threshold due to the persistent rainfall of November and, under the threat of breach, the discharge level was increased almost three times. The river Adyar, which originates from Chembarambakkam Lake, could not contain the extra discharge, leading to inundation of a large area of adjoining settlements, with Chennai city being completely submerged (Szynkowska, 2016) (see also Down to Earth, 2016).

Moving on to a holistic analysis of the rainfall aberration scenario, one can observe the complete Northeast Monsoon (October–December) occurrence of 2015. The district-wise estimates of precipitation (Table 6.9 in the appendix) in Tamil Nadu and Puducherry show a clear general trend of departure from observed normal rainfall, with departures exceeding 100%

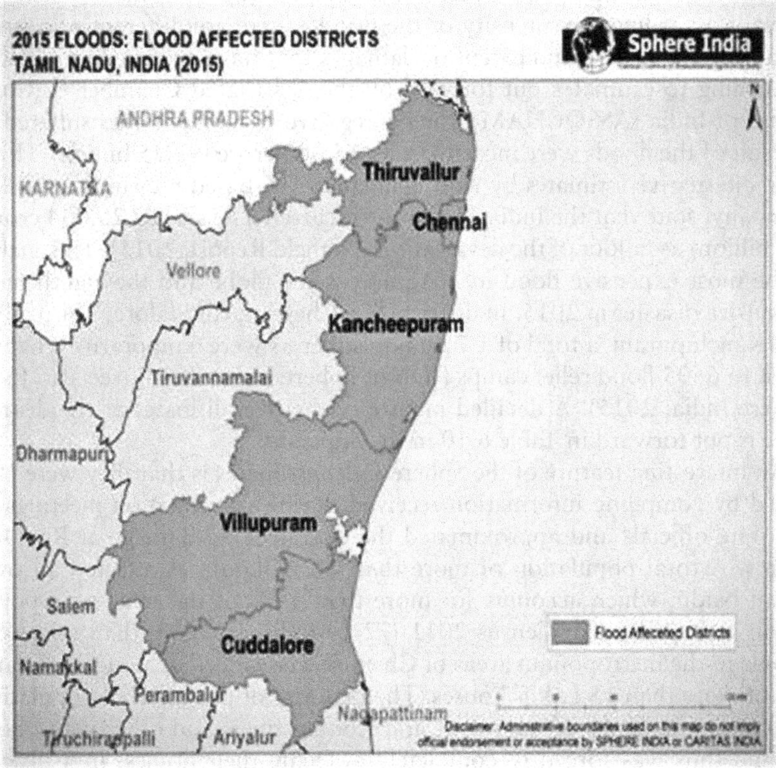

Figure 6.3 Flood-affected districts of Tamil Nadu in 2015
Source: Sphere India (2016)

for five districts. The interesting point derived here is that although Chennai received the least rainfall with respect to the other districts crossing the 100% mark, it was one of the worst hit as far as damage and displacements are concerned. Therefore, a perceivable feature is that, apart from the waterlogging from excess precipitation, the surfeit discharges of the rivers Adyar and Cooum were responsible for exacerbating the circumstances.

Damages and human displacement

As one would infer, the damage and losses as a result of the widespread flood across Tamil Nadu have been colossal. Since the majority of the urban poor reside near the banks of the Adyar and Cooum rivers, which are essentially low-lying regions, the floods proved a major cause of displacement, coupled with loss of shelter and basic amenities. Apart from this, coastal districts like Cuddalore and Thiruvallur incurred huge losses of population settlements and other infrastructural costs. Although a general consensus prevails regarding the intensity of the floods, there are different narratives regarding the nature and extent of damages that have occurred as a result. According to estimates put forward by the Associated Chamber of Commerce of India (ASSOCHAM), the aggregative economic losses suffered as a result of the floods were more than Rs 15,000 crore ($2.25 billion). However, alternative estimates by Aon Benfield, a UK-based reinsurance broker company, state that the Indian economy incurred a loss of Rs 20,034 crores ($3 billion) as fallout of the devastation (Benfield Report, 2015). This makes it the most expensive flood in 2015 across the globe and the eighth most expensive disaster in 2015. In districts like Chennai, Cuddalore, Thiruvallur and Kanchipuram, a total of 1.7 million sufferers were temporarily rehabilitated to 6605 flood relief camps (JDNA, Sphere India, 2016) (see also JNA, Sphere India, 2015). A detailed picture of the overall losses at the district level is put forward in Table 6.10 in the appendix.

An interesting feature of the Sphere India estimates is that they were prepared by compiling information received during coordination meetings of the state officials and approximated the overall flood damage at Rs 8481 crores. A total population of more than 10 million was affected all over Tamil Nadu, which accounts for more than 14% of the total state population as expressed in Census 2011 (72.14 million). More than 60 Lakhs people in the metropolitan areas of Chennai were affected, along with damage of more than 15 Lakhs houses. The majority of this affected population lived near the banks of the Adyar and Cooum rivers and other water bodies and thus was forced to completely evacuate their homes. In a district like Thiruvallur, there was severe damage to crops (24,870 hectares), with a significant impact on the overall food security of the district. Also, an estimate of about 15,000 hectares of land was announced to be completely inundated by the floodwaters, which propelled a displacement of a population of approximately 1.75 Lakhs. In terms of households, 6964 families lost their homes completely, thus plunging them into economic destitution, while 2925 families faced partial damage. However, the disastrous impacts on Cuddalore district have indeed been a cause of concern for the local administration and disaster relief workers. Almost 83% of the gram panchayats and 84% of the community blocks have been declared affected by the deluge. These together account for an affected population of over 6 Lakhs, with over 90,000 houses being damaged. Out of a total of 5409 huts, 837 were completely destroyed, while 2478 incurred partial destruction. Also, due to the overflow of the adjoining water channels, extensive damage to agricultural lands, settlements and infrastructure was recorded. Due to widespread inundations, several fields cultivating food crops faced immense damage across the district, thus escalating into a major food crisis. This also acted as a major push factor for the majority of the district's population to evacuate their homes and move to safer areas where they would not have to struggle in order to acquire food and basic necessities. In Villupuram district, major damage was incurred in the agricultural sector, with losses accounting for 3661 hectares of agricultural crops and 1548 hectares of horticulture. The damage incurred in 263 irrigation tanks proved a crucial setback in the district, coupled with impairment in 208 supply channels. The economic implication was therefore quite severe, with acute shortages of basic amenities like food, transportation, medical supplies and so on.

Apart from the material damage, there was also considerable delay in relief efforts in the initial days of the floods. It was reported that relief operations were carried out in a lopsided manner, and there seemed to be lack of coordination between the different organizations and agencies deployed for relief and rescue work. As a result, a substantial number of affected populations were marooned in a situation of abject deprivation and susceptibility during the catastrophic situation.

Rapid rate of urbanization as a reason behind the floods

One of the principal factors triggering the exacerbation of the Tamil Nadu floods is the rapid and unplanned urbanization that the state has been subject to for quite some time. Unplanned urbanization has emerged as a major problem in India, both at the micro and macro levels, and is a channel through which widespread environmental deterioration is taking place. Tamil Nadu is a rapidly growing state in India where the process of urbanization has been persistent for a fairly long time. Per Census 2011 estimates, a population of 34.92 million resides in the urban areas of Tamil Nadu as compared to 27.48 million in Census 2001. In terms of the share of urban to total population, Tamil Nadu holds the third position among the major states in India, with an estimated growth rate of 48.5%. A temporal analysis suggests that the level of urbanization in Tamil Nadu has been exhibiting a persistent positive trend across a century. The urban population has experienced an augmentation of more than four times in the last five decades, with a continuous increase in the number of towns. This phenomenon is depicted in Table 6.11 in the appendix and Figure 6.4.

A district-wise analysis of the urban expansion scenario (Table 6.12 in the appendix) suggests that Chennai is the only district in Tamil Nadu which is fully urban; that is, it has 100% rate of urbanization. Kanyakumari ranks second in this respect, with a 82.3% urbanization rate. The high urbanization rate of Kanyakumari district is intriguing, since it has no major industries under its belt and contradicts the urbanization around industries debate. Among the 32 districts, 15 have recorded estimates greater than the state level rate of 48.5%. However, one notable feature of this analysis is that apart from Villupuram, all the other flood-hit districts of Tamil Nadu have recorded considerable rates of urbanization. Districts like Thiruvallur (65.1%) and Kanchipuram (63.5%) have encountered a considerable share of urban expansion per Census 2011 estimates. This acts as a strong

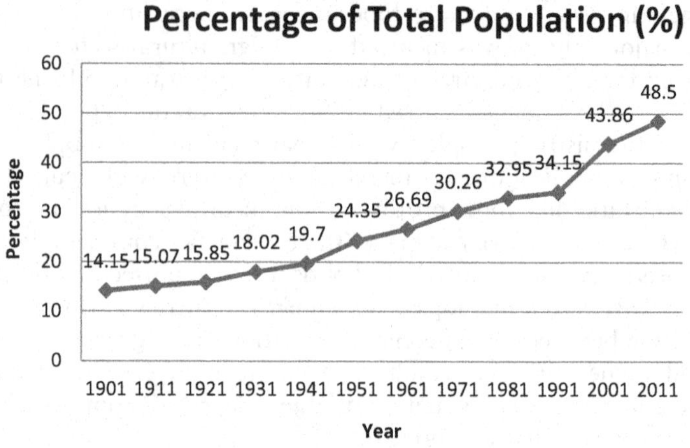

Figure 6.4 Trend in rate of urbanization in Tamil Nadu, 1901–2011

Source: Census of India, various years

supplement to the fact that uncontrolled urbanization stood out as a major cause of worsening the situation of the 2015 floods.

The economic implications of this rapid urban expansion are multifaceted. A crucial factor responsible for the increasing trend in urban augmentation is the gradual proliferation of increased population from rural areas into metropolitan areas. One of the primary motivations behind this surge in migration is securing appropriate employment opportunities and finding better facilities (Beine and Parsons, 2015). There is a prevalent gap between economic opportunities among the rural and urban areas of Tamil Nadu. Therefore, there is a strong economic incentive for a large quantum of the rural population to shift towards urban vicinities. According to Census 2011 estimates, 32.48% of all the rural migrants in Tamil Nadu listed their purpose of migration as related to work/employment. Moreover, Tamil Nadu is a state that experiences frequent droughts and monsoon failures. This has a palpable impact on the agricultural sector of the state, thus engendering a period of food insecurity and economic uncertainty. Therefore, this situation of economic crunch fosters the rate of migration from rural areas to urban spaces in Tamil Nadu. One of the problems due to such an influx of large numbers of migrants is the limited provision for quality and affordable housing. Due to the fixed availability of households and poor economic conditions, the surplus population has no choice but to settle in temporary dwellings and slum areas. As a result of such unprecedented mass relocation, tensions on the urban resources are created, leading to encroachment of urban greeneries and spaces near water bodies. This creates pressure on the urban services and infrastructure.

As a result of rapid urbanization, one can observe the disappearance of a number of water bodies and drainage systems that used to help in absorbing and reducing the intensity of floods. This has led to the burgeoning of impermeable concrete and asphalt spaces that act as strong deterrents to the trickling of water into the ground. As a result, the flood water fails to run off and therefore stagnates in low-lying areas. The water bodies and marshlands spread across the state have experienced rapid encroachments over time and presently house a large section of the underprivileged population, most of whom are migrants to urban areas. Apart from this, many wetlands have ceased to exist as a result of extensive construction work of residential buildings and commercial projects. As the city has experienced lateral expansion over time, the built-up area has also grown rapidly, jeopardizing the urban wetlands and water bodies. Moreover, the network of water bodies plays a crucial role in channelizing surplus water from one water body to another, which helps in maintaining a balance. Delving into the matter further, it is found that the initial encroachment is primarily through the buffer areas of a water body until the advent of the dry season, when the encroachment is rapid, thus wiping the water body completely off the map. The seasonal factor is particularly important, since during the dry season, due to stunted agricultural performance and other economic uncertainties, the rate of rural-urban migration increases, a reflection of which is seen in increased encroachment levels. In Chennai, the rapid rate of urban expansion with decimation of water sources was a major reason for its complete inundation and stagnation of flood water during the 2015 floods. Moreover, the areas that were centres of unplanned urbanisation across time were affected severely during the 2015 floods.

Thus, as more and more settlements come up in low-lying areas near water sources, the probability of devastation increases. Due to the overcrowding of these areas, urban services like sewerage and solid waste management are affected. As a result, water channels face blockages due to contamination of solid waste and therefore act as a major cause of submergence of adjoining low-elevation areas. Consequently, the underprivileged population residing in such areas is left with no choice but to go to rescue shelters or take refuge in safer places. Thus, one can see that the majority of the movers, as a result of the flood situation, are those who moved earlier into urban areas responding to economic stimuli like better occupational choice and basic amenities. Therefore, the case of the Tamil Nadu floods of 2015 revealed a transformation of a sizeable population from urban economic migrants to environmental migrants. This result has found importance in contemporary literature such as Szynkowska (2016).

Concluding remarks and policy prescriptions

It is prominently observed from this analysis that the problems pertaining to migration driven by climatic factors have multifaceted implications.

The degree of interplay of related factors and the degree of exposure of a population to the changing climatic conditions have been found to be crucial for human movement. Moreover, in India, the concept of climate change–induced migrants has been addressed at a minimal level with little or no data available to support the claims. As a result, it becomes exorbitantly difficult to draw the attention of policymakers and administrative representatives to these issues. In both cases cited in this analysis, the incompetence and negligence exhibited by administrative bodies had a major role in exacerbating the scenario. In the case of Malda, there still remains an urgent need to identify the vulnerable segments of the district population and build proper infrastructure to carry out rehabilitation initiatives. Although there have been considerable academic efforts to identify the points of departure of the river stream position of the Ganga and estimate the number of susceptible inhabitants, no such initiative has been taken by the district administration. As mentioned earlier, the situation needs to be envisaged from a more holistic perspective and not merely as a problem requiring disaster management initiatives in times of exigencies. Also, the escalated level of inland migration during flood situations and land cover loss brings to light a situation of tension between the migrants and the original inhabitants of a particular area. This needs to be dealt with at a structural level. Another point that needs to be acknowledged and acted upon is the uncertainty faced by the inhabitants of the char (channel bars). The majority of the population residing in these accreted areas of lands has been forced to move from their original settlements as a consequence of the land loss experienced in the left banks of the river Ganga. Due to the jurisdictional conflict between the state governments of West Bengal and Jharkhand, the fundamental rights of this population remain unaddressed. As for the case of the Tamil Nadu floods of 2015, two major inferences can be drawn from this chapter. First, there seem to a number of lacunae in the initiatives taken up by the governing bodies to restrict or prevent such severe floods. Across time, the phenomenon of urban flooding is becoming quite frequent in India, especially in the peninsular region. Moreover, Tamil Nadu, as a coastal state, always remains at risk of such precipitation anomalies due to depressions over the Bay of Bengal. Having said this, there have been very few initiatives to arrest the rapid rate of unplanned urbanization in the state. Also, the mass migration to urban vicinities from rural areas has worsened the situation many times over. Repetitive encroachments near water sources have wiped many out of existence and, as a result, during a deluge situation, the flood water tends to stall for days with no possibilities of smooth runoff. This dislocates a major population of underprivileged residents, and a transformation of identity occurs from economic to environmental migrants. Second, the relief initiatives in times of emergencies are often found to be lopsided, and as a result, many isolated groups of people remain trapped in the severe situation. In both cases cited, anthropogenic factors have been found to play a crucial role in exacerbating the problem of climate change. Many academicians argue that due to the innate heterogeneity in India's climatic conditions, it will emerge as a nation suffering substantially from the detrimental consequences of global climate change. Situated in such a condition, more attention needs to be devoted to the problem of human dislocation, since it is bound to escalate exponentially in the years to come and has the potential to consolidate into a national-level crisis.

Note

1 I would like to extend my sincere acknowledgements to Prof. Sunando Bandyopadhyay and Dr. Chalantika Laha Salui for granting me permission to use their findings regarding the change in stream position of the river Ganga across time. Valuable comments put forward by Dr Laha Salui and Dr. Kalyan Rudra have been of great help for this analysis. I would also like to express my sincere gratitude to Mr. S. Guchhait for granting me permission to use his computations and data regarding the area covered by the river Ganga across time in Manikchak and Kaliachak II blocks. I am also indebted to Dr. Architesh Panda for his valuable comments on the issue of climatic migration and its far-reaching implications.

References

Bates, Diane C. 'Environmental Refugees? Classifying Human Migrations Caused by Environmental Change.' *Population and Environment*, 23(5) (2002): 465–477.

Beine, Michel, and Parsons, Christopher. 'Climatic Factors as Determinants of International Migration.' *The Scandinavian Journal of Economics*, 117(2) (2015): 723–767.

Black, Richard. 'Environmental Refugees: Myth or Reality?' New Issues in Refugee Research, Research Paper No. 34, Geneva, UNHCR, 2001.

Black, Richard, Kniveton, Dominic, and Schmidt-Verkerk, Kerstin. 'Migration and Climate Change: Towards an Integrated Assessment of Sensitivity.' *Environment and Planning*, 43(2) (2011): 431–450.

Blaikie, Piers, Cannon, Terry, Davis, Ian, and Wisner, Ben. *At Risk: Natural Hazards, People Vulnerability and Disasters*. London: Routledge, 1994. https://doi.org/10.4324/9780203428764.

Castles, Stephen. 'Environmental Change and Forced Migration: Making Sense of the Debate.' New Issues in Refugee Research, No. 70, Geneva, United Nations High Commissioner for Refugees, 2002.

Census of India. 'Provisional Population Totals' (2013). Registrar General and Census Commissioner of India, Ministry of Home Affairs, New Delhi, India, 2011a.

Census of India. 'District Census Handbook: Malda', Village and Town wise Primary Census Abstract, Directorate of Census Operations, West Bengal, India, 2011b.

Census of India. Registrar General and Census Commissioner of India, Ministry of Home Affairs, New Delhi, India, 1901–2011.

Das, Tuhin K., Haldar, Sushil K., Gupta, Ivy Das, and Sen, Sayanti. 'River Bank Erosion Induced Human Displacement and Its Consequences.' *Living Reviews in Landscape Research*, 8(3) (2014).

Development and Planning Department, Government of West Bengal. 'District Human Development Report: Malda.' April 2007.

DNA. 'Chennai Flood Losses Estimated at Rs.20, 034 Crore to The India's Economy, Says Aon Benfield Report.' 2015. Available at: www.dnaindia.com/money/report-chennai-floods-losses-estimated-at-rs-20034-crore-to-the-india-s-economy-says-aon-benfield-report-2154521

Down to Earth. 'Chennai Apart.' 2016. Available at: www.downtoearth.org.in/news/urbanisation/chennai-apart-52265

El-Hinnawi, Essam. 'Environmental Refugees.' United Nations Environmental Program, Nairobi, 1985.

Falco, Chiara, Donzelli, Franco, and Olper, Alessandro. 'Climate Change, Agriculture and Migration: A Survey.' *Sustainability,* 10(2018): 1405.

Feng, Shuaizhang, Krueger, Allan B., and Oppenheimer, Michael. 'Linkages among Climate Change, Crop Yields and Mexico-US Cross-Border Migration.' *Proceedings of the National Academy of Sciences*, 107(32) (2010). https://doi.org/10.1073/pnas.1002632107.

Foresight Project and Government Office for Science, Great Britain. *Migration and Global Environmental Change: Future Challenges and Opportunities*. London: Government Office for Science, 2011.

Government of West Bengal. *Malda District Flood and Disaster Contingency Plan.* Government of West Bengal, India, 2017.

Guchhait, S. 'Quantification of River Bank Erosion, Accretion and Its Effect on Land Use: A Case Study of the Ganges (Left Bank) Upstream of Farakka Barrage, Malda District, West Bengal.' *Journal of Remote Sensing & GIS,* 9(1) (2018): 34–48.

Harris, John R., and Todaro, Michael P. 'Migration, Unemployment and Development: A Two-Sector Analysis.' *The American Economic Review*, 60(1) (1970): 126.142.

Homer-Dixon, Thomas F. *Environment, Scarcity and Violence*. Princeton, NJ: Princeton University Press, 1999.

Hugo, Graeme. 'Environmental Concerns and International Migration.' *International Migration Review*, 30(1) (1996): 105–131.

Indian Meteorological Department (IMD), Chennai. 'North East Monsoon Season.' 2015.Available at: www.imdchennai.gov.in/nemweb.pdf

Intergovernmental Panel on Climate Change (IPCC). 'Climate Change 2014: Impacts, Adaptation, and Vulnerability.' The Working Group II Contribution to the IPCC Fifth Assessment Report (WGII AR5), 2014a.

Intergovernmental Panel on Climate Change (IPCC). *Climate Change 2014: Synthesis Report. Contribution of Working Groups I, II and III to the Fifth Assessment Report of the Intergovernmental Panel on Climate Change* Core writing team, R. K. Pachauri, and L. A. Meyer (Eds.). Geneva: IPCC, 2014b.

Internal Disaster Monitoring Centre (IDMC). 'India.' 2017. Available at: www.internal-displacement.org/database/country/?iso3=IND

Keshkar, G., et al. 'Report of Experts Committee for Bank Erosion Problem of River Ganga-Padma in the Districts of Malda and Murshidabad.' *Planning Commission, Government of India* (1996): 1–71

Kibreab, Gaim. 'Environmental Causes and Impact of Refugee Movements: A Critique of the Current Debate.' *Disasters*, 21(1) (1997): 20–38.

Kondolf, Mathias G. 'Hungry Water: Effects of Dams and Gravel Mining on River Channels, Profile.' *Environmental Management,* 21(4) (1997): 533–551

Laha, Chalantika, and Bandyapadhyay, Sunando. 'Analysis of the Changing Morphometry of River Ganga, Shift Monitoring and Vulnerability Analysis using Space-Borne Techniques: A Statistical Approach.' *International Journal of Scientific and Research Publications*, 3(7) (2013).

Lewis, Arthur W. 'Economic Development with Unlimited Supplies of Labour.' *The Manchester School*, 22(2) (1954): 139–191

Majumdar, Samrat, and Mandal, Sujit. 'Channel Shifting of the River Ganga and Land Loss Induced Land Use Dynamicity in Diara Region of West Bengal, India: A Geo-Spatial Approach.' *International Journal of Basic and Applied Research*, 8(9) (2018).

Mandal, Sujit. 'Assessing the Instability and Shifting Character of the River Bank Ganga in Manikchak Diara of Malda District, West Bengal using Bank Erosion Hazard Index (BEHI), RS & GIS.' *European Journal of Geography*, 8(4:6) (2017).

Mondal Jayanta, Debanshi, Sandipta, and Mandal, Sujit. 'Dynamicity of the River Ganga and Bank Erosion Induced Land Loss in Manikchak Diara of Malda District of West Bengal, India: A RS and GIS Based Geo-Spatial Approach.' *International Journal of Applied Remote Sensing and GIS*, 3(1) (2016): 43–56

Munshi, Kaivan. 'Networks in the Modern Economy: Mexican Migrants in the US Labour Market.' *Quarterly Journal of Economics*, 118(2) (2003): 549–599.

Myers, Norman. 'Environmental Refugees in a Globally Warmed World.' *Bioscience*, 43(11) (1993): 752–761.

Myers, Norman. 'Environmentally-Induced Displacements: The State of the Art. Technical Report.' Environmentally-Induced Population Displacements and Environmental Impacts Resulting from Mass Migration, International Symposium, 1996.

Myers, Norman. 'Environmental Refugees.' *Population and Environment*, 19(2) (1997): 67–182.

Myers, Norman. 'Environmental Refugees: A Growing Phenomenon of the 21st Century', *Philosophical Transactions of the Royal Society* B, 357(1420) (2002): 609–613.

Myers, Norman, and Kent, Jennifer. *Environmental Exodus, an Emergent Crisis in the Global Arena*. Washington, DC: Climate Institute, 1995.

Narasimham, Balaji, Bhallamudi, S. Murty, Mondal, Arpita, Ghosh, Subimal, and Majumdar, Pradeep. 'Chennai Floods 2015 A Rapid Assessment.' Interdisciplinary Centre for Water Research Indian Institute of Science, Bangalore. 2016.

NDMA. 'National Disaster Management Guidelines – Management of Urban Flooding.' 2010.

Oliver-Smith, Anthony. 'Development and Dispossession: The Crisis of Forced Displacement and Resettlement.' Santa Fe, New Mexico: School for Advanced Research Press, 2009.

Panda, Architesh. 'Climate Refugees: Implications for India.' *Economic and Political Weekly* (2010): 76, 79.

Piguet, Etienne, Pecoud, Antoine, and Guchteneire, Paul de. 'Migration and Climate Change: An Overview.' *Refugee Survey Quarterly*, 30 (2011): 1–23.

Ravenstein, Ernst Georg. 'The Laws of Migration.' *Journal of the Royal Statistical Society*, 52(2) (1889): 241–305

Rudra, Kalyan 'The Farakka Barrage – An Interruption to Fluvial Regime.' *Indian Journal of Landscape Systems and Ecology Studies,* 19(2) (1996a): 105–110.

Rudra, Kalyan. 'Problems of River Bank Erosion along the Ganga in Murshidabad District of West Bengal.' *Indian Journal of Geography and Environment*, 1 (1996b): 25–32.

Rudra, Kalyan. 'Ganga Bhangan Katha: Malda-Murshidabad.' *Mrittika*, Kolkata, 2004a.

Rudra, Kalyan. 'The Encroaching Ganga and Social Conflicts: The Case of West Bengal, India.' Department of Geography, Habra S.C. Mahavidyalaya (College), West Bengal, India, 2004b.

Rudra, Kalyan. 'Dynamics of the Ganga in West Bengal, India (1764–2007): Implications for Science-Policy Interaction.' *Quaternary International*, 227(2) (2010): 161–169.

Ruiz, Vicente. 'Do Climatic Events Influence Internal Migration? Evidence from Mexico.' Working Papers 2017.19, FAERE – French Association of Environmental and Resource Economists, 2017.

Sinha, Rajiv, and Ghosh, Santosh. 'Understanding Dynamic of Large Rivers Aided by Satellite Remote Sensing: A Case Study from Lower Ganga Plains, India.' *Geocarto International*, 27(3) (2011): 207–219.

Soil and Land Use Survey of India. *Inventory of Soil Resources in Malda District, West Bengal Using Remote Sensing and GIS Techniques*. National Informatics Centre, Government of India, Kolkata, 2011.

Sphere India. *Joint Needs Assessment Report Tamil Nadu Floods-2015*. Inter Agency Group, Tamil Nadu and Sphere India, 2015.

Sphere India. *Tamil Nadu floods 2015 Joint Detailed Need Assessment Report – District Level: Shelter, Food, Nutrition & Livelihood and WASH*. 2016.

Stern, N. *The Economics of Climate Change: The Stern Review*. Cambridge: Cambridge University Press, 2007.

Szynkowska, Magdalena. '2015 Flood in Tamil Nadu, India, Disaster-Induced Displacement.' 2016. Available at: http://labos.ulg.ac.be/hugo/wp-content/uploads/sites/38/2017/11/The-State-of-Environmental-Migration-2016.152–172.pdf

Thakur, Praveen K., Laha, Chalantika, and Aggarwal, Shiv P. 'River Bank Erosion Hazard Study of River Ganga, Upstream of Farakka Barrage using Remote Sensing and GIS.' *Natural Hazards*, 61 (2011): 967–987.

Todaro, Michael P. *Internal Migration in Developing Countries*. Geneva: International Labour Office, 1976.

UNHCR. 'Frequently Asked Questions on Climate Change and Disaster Displacement.' 2016. Available at: www.unhcr.org/news/latest/2016/11/581f52dc4/frequently-asked-questions-climate-change-disaster-displacement.html

UNISDR, 'Annual Report 2017.' 2017. Available at: https://www.preventionweb.net/files/58158_unisdr2017annualreport.pdf

United Nations High Commissioner for Refugees (UNHCR). 'Convention and Protocol Relating to the Status of Refugees: Text of the 1951 Convention Relating to the Status of Refugees, Text of the 1967 Protocol Relating to the Status of Refugees, and Resolution 2198 (XXI) adopted by the United Nations General Assembly.' UNHCR, Geneva. United Nations, *Treaty Series*, 189 (2006): 137.

Zelinsky, Wilbur. 'The Hypothesis of the Mobility Transition.' *Geographical Review*, 61(2) (1971): 219–249.

Appendix

Table 6.1 Administrative information of Malda District

Administrative unit	*Number*	*Name*
Sub-division	2	Malda Sadar and Chanchal
Community development block	15	Bamangola, Chanchal-I, Chanchal-II, English Bazar, Gajole, Habibpur, Harischandrapur-I, Harischandrapur-II, Kaliachak-I, Kaliachak-II, Kaliachak-III, Malda (Old), Manikchak, Ratua-I, Ratua-II
Gram Panchayat	146	
Number of mouzas	1814	
Number of villages	3701	
Municipality	2	English Bazar and Old Malda
Headquarters		English Bazar

Source: Census 2011

Table 6.2 Areas under different soil erosion classes

Erosion classes	*Area (ha)*	*Area (%)*
None to slight	151,294	40.53
Slight to Moderate	113,362	30.37
Moderate	79936	21.41
Moderate to Severe	1036	0.28
Misc.	27672	7.41
Total	373,300	100

Source: Soil and Land Use Survey of India, 2011

Table 6.3 Block profile of Manikchak

Block variables	*Number*
Geographical area (in sq km)	209.2
No. of mouzas	89
No. of villages	104
Gram Panchayats	11
Total population (Census 2011)	2,69,813
Male population	1,39,593
Female population	1,30,220
Scheduled caste population	74,816
Scheduled tribe population	40,125
No. of primary schools	150
No. of secondary schools	8
No. of higher secondary schools	13
No. of colleges	Nil
No. of primary health centres	1
No. of beds	25
No. of secondary health centres	35
No. of post offices	23
No. of bank branches	10

Source: Census 2011

Table 6.4 Area covered by Ganga and the number of affected mouzas in Manikchak block

Year	*Area covered by Ganga*			*Number of affected mouzas*
	Less than 30%	*30%–60%*	*More than 60%*	
1980	Dergama, Chandpur Tafir, Mirpur, Manikchak, Rahimpur, Dakshin Chandipur, Chandipumal, Bagdukra, Hiranandapur, Gopalpur	Masaha, Narayanpur, Gadai, Paschim Narayanpur, Rostampur, Duani Tafir	Raniganj, Gobindapur, Rambari	19
1990	Rostampur, Gobindapur, Paschim Narayanpur, Duani Tafir, Raniganj, Chandipur Tafir, Kesarpur, Chandipumal, Bagdukra, Sobhanathpur, Dharamapur, Dakshin Chandipur, Sukhsena, Gopalpur	Manikchak, Hiranandapur, Narayanpur, Gadai	Rahimpur, Mirpur, Rambari, Dergama, Masaha	23

Year	*Area covered by Ganga*			*Number of affected mouzas*
	Less than 30%	*30%–60%*	*More than 60%*	
2000	Rahimpur, Sukhsena, Jot Bhabani, Kesarpur, Chandipumal, Chandipur Tafir, Rostampur, Gobindapur	Dharampur, Rambari, Gadai, Narayanpur, Manikchak, Dakshin Chandipur, Duani Tafir, Paschim Narayanpur	Dergama, Mirpur, Hiranandapur, Masaha, Sobhanathpur, Gopalpur, Bagdukra	23
2010	Sukhsena, Bagdukra, Hiranandapur, Chandipur Tafir, Narayanpur, Kesarpur, Duani Tafir, Dergama, Chandipumal	Rahimpur, Dharampur, Paschim Narayanpur, Rambari, Manikchak, Dakshin Chandipur, Mirpur, Gadai, Gopalpur	Masaha, Sobhanathpur	20
2017	Gadai, Rahimpur, Naobarar Jaigir, Mirpur, Chandipur Tafir, Jot Bhabani, Chandipumal, Rostampur	Masaha, Kesarpur, Narayanpur, Sukhsena, Paschim Narayanpur, Dakshin Chandipur, Rambari, Gopalpur	Sobhanathpur, Duani Tafir, Dharampur, Manikchak, Bagdukra, Dergama, Hiranandapur	23

Source*:* Guchhait (2018)

Table 6.5 Affected Gram Panchayats with number of affected population in Manikchak, 2016–17

Name of affected Gram Panchayat	*Displaced population*
Dakshin Chandipur	30,000
Manikchak	2000
Gopalpur	1500
Dharampur	1500
Hiranandapur	4000
Mathurapur	1500

Source*:* Disaster Management Plan of Malda district, 2017

Table 6.6 Block profile of Kaliachak II

Block variables	*Number*
Geographical area (in sq km)	209.17
No. of mouzas	43
No. of villages	317
Gram Panchayats	9
Total population (Census 2011)	2,10,105
Male population	1,07,553
Female population	1,02,552
Scheduled caste population	32,686
Scheduled tribe population	4,816
No. of primary schools	127
No. of secondary schools	11
No. of higher secondary schools	13
No. of colleges	Nil
No. of primary health centres	2
No. of beds	30
No. of secondary health centres	34
No. of post offices	16
No. of bank branches	11

Source: Census, 2011

Table 6.7 Area covered by Ganga and the number of affected mouzas in Kaliachak II block

Year	*Area covered by the Ganges*			*Number of affected mouzas*
	Less than 30%	*30%–60%*	*More than 60%*	
1980	Kankribandha Jhaubona, Birodhi, Kamaluddinpur, Mahadebpur, Jotkasturi, Paranpur, Rezakpur	Char Babupur, Daridiar Jhaubona, Hamidpur, Panchanandapur, Nayagram	Daskathia, Gaziapara, Darijayrampur, Dogachhi	16
1990	Birodhi, Daskathia, Dogachhi, Jotkasturi, Panchanandapur, Darijayrampur, Paranpur	Nayagram, Kamaluddinpur, Gaziapara, Char Babupur, Mahedebpur, Hamidpur	DaridiarJhaubona, Islampur, Kankribandha Jhaubona	16

Year	Area covered by the Ganges			Number of affected mouzas
	Less than 30%	30%–60%	More than 60%	
2000	Hamidpur, Jotkasturi, Panchanandapur, Daridiar Jhaubona, Birodhi, Darijayrampur, Dogachhi, Jot Ananta	Nayagram, Mahadebpur, Kamaluddinpur, Paranpur, Kankribandha Jhaubona, Gaziapara, Daskathia	Islampur, Char Babupur, Rezakpur	18
2010	Jotkasturi, Paranpur, Gaziapara, Shukurullapur, Mahadebpur, Daridiar Jhaubona, Nayagram, Kamaluddinpur, Panchanandapur, Char Babupur, Jot Ananta, Hamidpur, Darijayrampur	Birodhi, Kankribandha Jhaubona, Rezakpur	Dogachhi	17
2017	Rezakpur, Mahadebpur, Nayagram, Shukurullapur, Jotkasturi, Panchanandapur, Jot Ananta, Hamidpur, Darijayrampur	Char Babupur, Kamaluddinpur, Kankribandha Jhaubona, Birodhi	Paranpur, Daridiar Jhaubona, Dogachhi, Gaziapara	17

Source: Guchhait (2018)

Table 6.8 Affected Gram Panchayats with affected population in Kaliachak II, 2016–17

Name of affected Gram Panchayat	*Displaced population*
Bangitola	3000
Hamidpur	1500
Rajnagar	2000
Panchanandapur I & II	3500

Source: Disaster Management Plan of Malda district, 2017

Table 6.9 District-wise precipitation in Tamil Nadu and Puducherry (October to December 2015)

District	*Actual rainfall (mm)*	*Normal rainfall (mm)*	*Departure (%)*
Ariyallur	681.2	545.5	24.8
Chennai	1608.6	789.9	103.6
Coimbatore	341.1	328.9	3.7
Cuddalore	1240.5	697.8	77.8
Dharmapuri	449.6	330.1	36.2
Dindigul	486	436.4	11.4
Erode	396.8	314.6	26.1
Kanchipuram	1815	641.8	182.8
Kanyakumari	780.1	496.4	57.2
Karaikal	1316	1048.5	25.5
Karur	360.5	314.7	14.5
Krishnagiri	442.4	289.4	52.9
Madurai	413.1	419.1	-1.4
Nagapattinam	1378.8	941	46.5
Namakkal	308.4	291.6	5.8
Nilgiris	565.7	478.2	18.3
Perambalur	525.8	440.9	19.3
Puducherry	1556.6	843.1	84.6
Pudukottai	573.6	406.2	41.2
Ramanathapuram	572.4	491.7	16.4
Salem	488	370.5	31.7
Sivagangai	460.1	422.7	8.8
Thanjavur	693.5	550.3	26
Theni	399.4	357.9	11.6
Tirunelveli	1050.6	467.2	124.9
Tiruppur	338.8	314.3	7.8
Thiruvallur	1466.6	589.3	148.8
Tiruvannamalai	595.8	446.5	33.4
Tiruvarur	1022.3	719.1	42.1
Toothukudi	664.3	427	55.6
Tiruchirapalli	440.8	391.5	12.6
Vellore	747.7	348.7	114.4
Villupuram	928.1	499.1	85.9

Source*:* Indian Meteorological Department (IMD), Chennai

Table 6.10 District-wise assessment of losses and human displacement as a result of floods in 2015

Tamil Nadu (state highlights):	
Total population of state	72.14 million
Total population affected in state	More than 10 million
Affected houses	More than 25 Lakhs
Flood damages	Rs. 8481 Crore

Tamil Nadu (state highlights):	
Chennai urban	
Total population	4,646,732
Population affected in Chennai	More than 60 Lakhs
Houses affected	More than 15 Lakhs
Thiruvallur district-	
Total population	3,728,104
Population affected	Approx. 1.75 Lakhs
Houses affected	More than 51,000
Total no. of households	946,949
Total number of families who lost their houses (fully damaged)	6964
Total number of families whose houses are partially damaged	2925
Total number of livestock lost	2218
Crop damaged in hectares	24,870
Land submerged in water in hectares	15,000
Kanchipuram district:	
Total population	3,998,252
Affected population	More than 10 Lakhs
Houses affected	More than 1.9 Lakhs
Cuddalore district:	
Total population	2,605,914
Affected population	More than 6 Lakhs
Houses affected	More than 90,000
GPs affected	500 GPs affected out of 600
Blocks affected	11 blocks affected out of 13
Total no. of huts	5409
Huts damaged	Fully damaged: 837 Partially damaged 2478
Villupuram district:	
Total population	3,458,873
Population affected	More than 80,000
Houses affected	Approx. 20,000
Blocks affected	22
Houses damaged	Fully: 971 (local NGO information) Partially: 15,204
Cattle killed	2442
Crops damaged in hectares	3661 agriculture 1548 horticulture
Irrigation tanks damaged	263
Supply channels damaged	208

Source: Sphere India, 2016

Table 6.11 Urbanization in Tamil Nadu, 1901–2011

Year	*No. of towns*	*Urban population (millions)*	*Percentage of total population (%)*	*Decadal growth (%)*
1901	133	2.72	14.15	–
1911	162	3.15	15.07	15.81
1921	189	3.43	15.85	8.89
1931	222	4.23	18.02	23.32
1941	257	5.17	19.7	22.22
1951	297	7.33	24.35	41.78
1961	339	8.99	26.69	22.65
1971	439	12.46	30.26	38.60
1981	434	15.95	32.95	28.01
1991	469	19.08	34.15	19.62
2001	832	27.48	43.86	44.03
2011	1097	34.92	48.5	27.07

Source: Census of India, various years

Table 6.12 District-wise estimates of rate of urbanization in Tamil Nadu

District	*Rate of urbanization (%)*
Chennai	100
Kanyakumari	82.3
Coimbatore	75.7
Thiruvallur	65.1
Kanchipuram	63.5
Tiruppur	61.4
Madurai	60.8
The Nilgiris	59.2
Theni	53.8
Erode	51.4
Salem	51
Virudhunagar	50.5
Tuticorin	50.1
Tirunelveli	49.4
Tiruchirappalli	49.2
Vellore	43.2
Karur	40.8
Namakkal	40.3
Dindigul	37.4
Thanjavur	35.4
Cuddalore	34
Sivagangai	30.8
Ramanathapuram	30.3
Krishnagiri	22.8
Nagapattinam	22.6

District	*Rate of urbanization (%)*
Thiruvarur	20.4
Thiruvannamalai	20.1
Pudukkottai	19.5
Dharmapuri	17.3
Perambalur	17.2
Villupuram	15
Ariyalur	11.1
State	48.5

Source*:* Census, 2011a

Part III

Responses to climate change from South Asian nations

7 India's commitment to counter climate change in South Asia

A critical evaluation

Jayita Mukhopadhyay

Amidst worldwide concern about the unsettling impact of climate change on human civilization, and in the middle of the cauldron created by struggle of the countries of South Asia to fulfil their commitments to the Paris Climate Agreement within the United Nations Framework Convention on Climate Change (UNFCCC), India is becoming the focal point of attention, and its response is being keenly watched and analysed. India is not only the largest and most populous country in South Asia, it is also the most resource-rich one, as one of the world's fastest-growing economies, and according to many estimates the sixth-largest economy of the world, with a nominal GDP of $2.61 trillion. What is even more compelling, India is world's largest functional, thriving democracy, loudly proclaiming its commitment to hallowed norms of liberty, equality, and justice not only at home but also in the international arena. India harbours a deep-rooted ambition to secure the high pedestal of a regional superpower, of becoming a rule maker rather than a rule taker. So the onus of reversing climate change, particularly of an anthropogenic nature, mitigating the adverse impact of climate change in South Asia, promoting adaptation strategies to climate change, providing leadership to other countries of the region in this regard, coordinating their activities, and acting as the interlocutor between regional interests and global commitments in climate change–related issues squarely rests on India. As a regional behemoth, it has a huge responsibility towards its South Asian neighbours threatened by climate change. Bangladesh and the island country Maldives face the danger of being submerged in future if climate change is not responded to adequately, and the threats faced by other countries of South Asia are equally palpable. India's task is made more challenging by the fact that India simultaneously faces the challenge of lifting about 276 million people above the poverty line, according to various government sources, and meeting the rising demand of the other part of its mammoth population for better infrastructures and more intense urbanization and industrialization.[1] Hence, it can't significantly compromise its development agenda. The political economy of India creates overriding pressures on our policy-makers to continuously seek a point of equilibrium between resource mobilization and sustainability.

Objective of the chapter

This chapter begins with a brief enumeration of the major kinds of climate change–related threats faced by India and then proceeds to probe how it can cope with these challenges at home and at a larger, regional level. It explores and analyses what our politicians, particularly policy-makers, are presently doing to deal with climate change within the country, whether they are lacking in resolve, and what else needs to be done. It then proceeds to examine how far India has been successful in utilizing regional bodies meant for promoting intercountry cooperation, particularly the South Asian Association for Regional Cooperation (SAARC) in promoting environmental activism, and what can it do to further bolster such efforts. India's activist role in the global battle against climate change is also examined briefly. The chapter intends to produce some useful perspectives on how to avert the apocalyptic consequences of climate change in South Asia.

Climate change scenario in India

India signed the Paris agreement in November 2017, which seeks to limit the Earth's warming to below 2 degrees Celsius. But to keep its pledge, India needs to confront the unpalatable reality that it alone accounts for 4.5 per cent of the world's greenhouse gases. India's carbon emission is still rising, as it is heavily dependent on fossil fuel to meet its energy needs, which, again, are exponential, given India's drive towards poverty eradication, industrialization, and urbanization. Data revealed by an important study done by Global Carbon Project on the eve of UN Climate Change Conference (COP 24) in Katowice, Poland, in December 2018, indicates that India is now the third-largest contributor to carbon emission, superseded by China and the United States, respectively, in first and second position, a dubious distinction which India would definitely like to dispense with.[2] Thus, India is a part of the problem, as it is a major contributing party in climate change. But India is also a victim of the problem and, most promisingly, a country keen on and capable of providing the solution to the problem.

India is a victim of various adverse fallouts of climate change initiated by reckless resource mobilization and exploitation of mother earth by today's developing countries through centuries who have now reached comfortable levels of economic growth and can consequently afford costly clean energy and modern technology so as to reduce their carbon emission levels. They can now pontificate to the developing countries about their duty of cutting down emissions while ignoring the development imperatives of these emerging economies. Like other South Asian countries, India is witnessing frequent changes in rainfall patterns, fluctuating weather conditions, and extreme drought alternating with flash floods, all of which are contributing to poor agricultural production, thus aggravating food scarcity, accentuating poverty, and triggering internal and cross-border migration, leading to excess pressure on our cities. One problem is leading to another, and climate change is thus setting in motion a cycle of poverty, underemployment, malnutrition, declining standards of life, gender inequality, destruction of village economy, urban squalor and poverty, crumbling law and order situation, psychosomatic distress, and myriad other dystopian outcomes. The Economic Survey Report of India 2018, produced under the tutelage of then-Chief Economic Advisor Arvind Subramanian and his team, made some grim observations on the fate of Indian farmers due to the adverse impact of climate change. The report noted that around 52 per cent of India's cultivable land is still unirrigated and rain fed and bears the brunt of temperature variation and erratic rainfall. Climate change models developed by the Intergovernmental Panel on Climate Change (IPCC) predict that by the end of the 21st century, temperatures in India are likely to rise by between 3 and 4 degrees Celsius. Given the low capacity of Indian farmers to adapt to climate conditions, the survey made the grim prediction that farm incomes will be lower by around 12 per cent on an average in the coming years, particularly in unirrigated areas and annual revenue may experience a potential losses amounting to almost 18 per cent.[3] Agriculture being the backbone of the Indian economy, such predictions do not augur well for the overall wellbeing of the country and its population and underscore the urgency of addressing climate change in earnest. A steady and quite alarming rise in temperatures, erratic rainfall patterns, droughts and severe water scarcity in some pockets of the country hampering agriculture, devastating storms, cataclysmic deluges, frequent flash floods, wildfires, landslides, rising sea levels, accelerated erosion of coastal zones, species extinction, and the spread of vector-borne diseases are some of the major adverse manifestations of climate change which India has witnessed quite frequently over the last decade or so. Celebrated Indian author Amitav Ghosh, in his magnum opus, *The Great Derangement*, has presented his brilliant analysis of the pervading impact of climate change on the lives of people in the Indian subcontinent against the backdrop of an enumeration of some notable extreme climatic events in India in the recent past whose intensities and frequencies have increased steadily (Ghosh, 2016). A brief recounting of some major events of the last decade can put things in perspective. In July 2005, unprecedentedly heavy rainfall virtually paralysed the city of Mumbai, India's commercial capital, causing loss of life and severe damages to property and totally unnerving the confidence of the inhabitants of a scintillating urban landscape normally bustling with activity. The idyllic landscape of Ladakh, in the trans-Himalayan region, the favourite location of film directors and ideal vacation place of tourists, was ravaged by a severe flash flood caused by cloudburst in August 2010, leading to widespread destruction of life and property and causing geomorphologic changes. Similar incidents also occurred there in 2015, thereby underscoring the increasing fragility of the Himalayan region, caused by rising temperatures, shifting precipitation trends, and other adverse outcomes of reckless human action. Similar incidents occurred in Uttarakhand in June 2013 and in Jammu and Kashmir in September 2014. A large part of the city of Chennai was inundated in November 2015 due to exceptionally heavy rainfall caused by a retreating southwest monsoon whose severity was unmatched. The southern state of Kerala (popularly famed as God's own country) perhaps incurred God's wrath in August 2018, when a flood wreaked havoc, the worst to hit the state in a hundred years. As aid from domestic and international sources poured in, and the people of Kerala fought a valiant battle to get back on their feet, environmentalists again sounded the alarm bell to drive home the message that climate change can be ignored only at our peril.

Initiatives at domestic level to deal with climate change

Fortunately for India, unlike some Western leaders like present US President Donald Trump, India's political establishment has not denied climate change and its many challenges. Over the years, those at the helm of governance have put in place a National Action Plan and various regulatory frameworks and institutional arrangements to deal with climate change, though opinions are sharply divided about the effective implementation of these plans and mechanisms and the sincerity and resolve of our policymakers. Ever since the Rio Earth Summit 1992, India shared the concern of the global community about the urgency of tackling climate change and mitigating its adverse effects, but it was unwilling to make any commitments regarding cutting down greenhouse gas emissions, as it prioritized its developmental needs and energy demands. It insisted that agenda of cutting down greenhouse gas emissions must be based on concepts of common but differentiated responsibilities and respective capabilities. In other words, it advocated the doctrine of climate justice and urged developed countries to own up to their historic misdeeds of colonial exploitation and rampant, reckless use of natural resources to fuel their industrial growth, thereby creating the problem of climate change in the first place and now, as a compensatory measure, to provide financial assistance to developing countries so that they may use clean energy and better technology which will save the planet without compromising their development agenda. Thus, concern for equity, anxiety about the fairness of international climate negotiations, and a strong aversion to outside forces dictating domestic policies prevented India from making any policy at the national level to deal with climate change (Atteridge et al. 2012). But in the post-liberalization era, as India's economy grew from strength to strength and its technological abilities, particularly those used in tapping renewable sources of energy like solar, hydro power, and wind power made rapid strides, India changed its official stance towards climate change both domestically and in the international sphere. International factors like anxiety to attain the status of a global actor; fear of isolation as another prominent polluter, China, proceeded to accept limits on its carbon emissions; and domestic factors like the increasing assertiveness of civil rights groups demanding reductions in carbon emissions made India take noticeable steps towards the goal of mitigating climate change, starting from 2007, continuing to the present.

However, developments in this period were preceded by slow but steady measures adopted by our leaders to counter the menace of climate change. Indira Gandhi was the first prime minister of India to take matters related to environmental degradation and climate change very seriously back in 1970s. She was the only head of a government to attend the United Nations Conference on the Human Environment, also known as Stockholm Conference, held in 1972, and, at her behest, the Air Pollution Control Act, Wildlife Protection Act, Water Pollution Control Act, and Forest Conservation Act were implemented, thereby generating significant awareness among the general public and civil rights groups about the need to deal with climate change. Mrs. Gandhi's initiatives gave birth to new ideas and activism and greatly shaped India's subsequent policy on climate change (Rangarajan, 2007).

The next round of significant moves made by our policy-makers happened, as already noted, in 2007. Prime Minister Dr. Manmohan Singh announced in 2007 that India's per capita emissions would never exceed those of industrialized countries, thereby signalling India's voluntary move towards restricting carbon emissions. A watershed development took place in 2008 with the announcement of India's National Action Plan on Climate Change (NAPCC). A plethora of significant steps to simultaneously advance India's development and climate change–related objectives were laid down by the National Action Plan. The plan initially established eight national missions to address various aspects of climate change mitigation and adaptation.[4] Let us focus on these eight missions.

The *National Solar Mission* was introduced with the objective of promoting the development and use of solar energy for power generation and other uses. The mission initiated the establishment of a solar research centre, increased international collaboration on technology development, strengthened domestic manufacturing capacity, and increased government funding and international support. Because of the big push for solar energy, solar panels have now been installed in many public and even private institutes, solar lanterns have become more affordable, and all these measures are now causing significant reduction in the use of fossil fuels. About 7 per cent of India's installed power capacity now comes from solar power, which bears testimony to the success of this mission.

The *National Mission for Enhanced Energy Efficiency* encouraged large energy-consuming industries to adopt more sophisticated, state-of-the-art technologies which are more energy efficient and incentivized the switchovers by offering tax exemptions and energy-saving certificates. Specific demand-side management programs in the municipal buildings and agricultural sectors were also undertaken to minimize use of fossil fuel, and all these measures have produced dividends by promoting energy efficiency.

The *National Mission on Sustainable Habitat* began with the objective of promoting energy efficiency as a core component of urban planning by extending the existing Energy Conservation Building Code, strengthening the enforcement of automotive fuel economy standards, and using pricing measures to encourage the purchase of efficient vehicles and incentives for the use of public transportation. The NAPCC also laid stress on waste management and recycling and has been reasonably successful.

The *National Water Mission*, in collaboration with Asian Development Bank, aimed at achieving a 20 per cent improvement in water use efficiency through pricing and other measures to cope with water scarcity caused by climate change.

The *National Mission for Sustaining the Himalayan Ecosystem* aimed at preventing the melting of the Himalayan glaciers and protecting biodiversity in the Himalayan region and has been moderately successful. Though the mission is in place, unplanned construction in the upper edges of Uttarakhand and adjoining areas have continued unabated. This has accentuated the threat of landslides and flash floods. Critics have blamed the unholy nexus between greedy promoters and local politicians and government officials for this undesirable state of affairs.

The *Green India Mission*, aimed at afforestation of 6 million hectares of degraded forest lands and expanding forest cover from 23 to 33 per cent of India's territory, has made good progress in achieving its goal. It is making various efforts to enhance annual carbon sequestration by 50 to 60 million tonnes in the year 2020 through promotion of organic farming and other eco-friendly practices.

The *National Mission for Sustainable Agriculture*, aimed to support climate adaptation in agriculture through the development of climate-resilient crops, expansion of weather insurance mechanisms, and healthy agricultural practices, also has been moderately successful.

The *National Mission on Strategic Knowledge for Climate Change*, aimed at networking existing knowledge institutions, capacity building, and improving understanding of key climate processes and climate risks, has continuously taken many steps to educate citizens about the danger of climate change and sensitize them to ways and means of mitigating climate change. It has also encouraged private-sector initiatives to develop adaptation and mitigation technologies through venture capital funds. It has facilitated use of technology in specific regions for efficient use of non-renewable sources of energy.

Various other significant initiatives have been undertaken by government under NAPCC.

The government is mandating the retirement of inefficient coal-fired power plants and supporting the research and development of Integrated Gasification Combined Cycle (IGCC) and supercritical technologies. Under the Energy Conservation Act 2001, large energy-consuming industries are required to undertake energy audits, and an energy-labelling program for appliances has been introduced. Under the Electricity Act 2003 and the

National Tariff Policy 2006, the central and the state electricity regulatory commissions must purchase a certain percentage of grid-based power from renewable sources. India is fortunate to be endowed with renewable sources of energy like sunlight, biomass, wind, and water, and the government has made a consistent push to the private and public sectors to tap and utilize these resources. According to government estimates, energy from renewable sources now accounts for about 33 per cent of India's primary energy consumptions. India is increasingly adopting responsible renewable energy techniques and taking positive steps towards reducing carbon emissions, cleaning the air, and ensuring a more sustainable future, and for this, our policy-makers, government employees, scientists, researchers, technocrats, and bureaucrats deserve appreciation. But, needless to say, much remains to be done.

The NAPCC also has as an integral part a specific regulatory mechanism. Ministries with lead responsibility for each of the missions are directed to develop objectives, implementation strategies, timelines, and monitoring and evaluation criteria to be submitted to the Prime Minister's Council on Climate Change. The Council has been entrusted with the task of periodically reviewing and reporting on each mission's progress.

Like all policy initiatives taken by the government in India, the NAPCC has received both bouquets and brickbats. Noted environmentalist Sunita Narain lauded the Action Plan, as she felt that the plan was an acknowledgement on India's part that it is willing to achieve its developmental agenda in a different, responsible way. Besides, the plan prioritized national action over regional initiatives, thereby accepting the pervasive nature of climate change. But another noted researcher and water expert, Sudhirendar Sharma, has dismissed the plan report as a compilation of listless ideas that lack depth, vision, and urgency. Others have castigated it for not taking specific care of the needs of the most vulnerable, poorest sections of people who are bound to depend on nature for their survival (Pandve, 2019).

Though opinion is sharply divided about the efficacy of the Plan introduced in 2008 and pursued to date, there is genuine appreciation expressed by people of different walks of life for some very significant moves made by Jairam Ramesh as minister for environment and forests in 2009. Ever since his appointment as India's environment minister, Ramesh has launched many initiatives to create awareness and promote activism among ordinary citizens as well as captains of industries in our country to take appropriate measures to mitigate the impact of climate change. Under his leadership, India became more proactively engaged in climate change negotiations internationally as well.

SAARC initiatives and India's role in dealing with climate change

As one of the most prominent founding members of the South Asian Association for Regional Cooperation, which began its journey in 1985 (Dhaka Summit), India, the most powerful economy of the region and the sole

SAARC country sharing its borders with all six members via land and sea (Bangladesh, Bhutan, Pakistan, Nepal, Maldives, and Sri Lanka), was naturally expected to play a leading role in its deliberations and regional initiatives. India has tried to fulfil expectations by making major financial contributions for the promotion of welfare measures in the member countries, which also includes climate-related initiatives. But unfortunately, some political issues, particularly lingering tension between India and Pakistan and China's persistent effort to bring other South Asian countries into the ambit of its influence by offering them credits and investments with the objective of isolating and stymieing India, has adversely effected concerted efforts by SAARC countries to manage climate change–related issues. Such political bickering has thwarted speedy development of a Regional Environmental Governance Mechanism, though South Asia needs it urgently. The Intergovernmental Panel on Climate Change has predicted a grim future for South Asia. It has observed that the frequency of hot days is likely to increase, as revealed by study of temperature trends; heavy rainfall days and intense dry days are also likely to happen more, and rising sea level is likely to threaten South Asia's coastal settlements, coastal economies, cultures, and ecosystems, and the threat may be aggravated by cyclone frequency or intensity. South Asia being an impoverished region with an average gross national income (GNI) per capita of approximately US$1,610 (2016 figure), climate change is ominous for the region, as the relation between poverty and environmental degradation has been firmly established. India already faces the problem of tackling environmental refugees from neighbouring states and spread of vector-borne and other communicable diseases triggered by climate change. India has major river water sharing disputes with Pakistan, Bangladesh, Nepal, and other neighbours, which are, again, irritants to collective effort (Zafarullah and Huque, 2018).

Nevertheless, the SAARC did address environmental concerns. In the Kathmandu Declaration, 1987, issued at the Third SAARC Summit, all the heads of states of SAARC expressed 'their deep concern' for the regional challenges related to environmental degradation and climate change. They also recognized that these challenges are 'severely undermining the development process and prospects of the member countries', and they 'decided to intensify regional cooperation with a view to strengthening their disaster management capabilities'.[5]

To realize the declared objective, SAARC commissioned a study on the Protection and Preservation of the Environment and the Causes and Consequences of Natural Disasters, which was finalized in 1991. In 1992, the Technical Committee on Environment was established and was entrusted with the task of identifying measures for immediate action and deciding modalities for implementation. The scope of the committee was gradually expanded to address emerging issues. However, little attention was paid to recommended measures, and because of an absence of political will, no collective step could be taken.

In several subsequent declarations, SAARC has expressed a concern for environmental issues, including climate change. Only after the devastating Indian Ocean tsunami (2004), in 2005, did SAARC members agree on concrete actions to address natural disasters and climate-related disasters. The Comprehensive Framework on Disaster Management adopted by SAARC in 2006 led to the creation of the SAARC Disaster Management Centre (SDMC) in October 2006, which focused on advising members about climate policy and facilitated capacity building. But lack of unity and commitment among members to make SAARC a really effective and integrated regional body prevented the implementation of many well-conceived ideas.

The SAARC Convention on Cooperation on Environment, adopted in Thimpu at the sixteenth SAARC Summit (2010) and ratified by member states in 2013, once again tried to energize collective efforts of member states in mitigating climate change. SAARC established an Expert Group on Climate Change to ensure policy direction and guidance for regional cooperation. Most SAARC declarations acknowledged the role of the UN Framework Convention on Climate Change and the National Adaptation Programmes of Action (NAPAs) as the two pillars of climate action. At Thimpu, decisions were made to establish an Inter-Governmental Monsoon Initiative to study the evolving pattern of monsoons in the region so as to assess the region's vulnerability to climate change. The Inter-Governmental Mountain Initiative was set up to study mountain ecosystems and to analyse their impact on people's livelihood. The Inter-Governmental Marine Initiative was set up to bolster understanding of shared oceans and water bodies in the region and to provide insight about how to use them judiciously so as to promote sustainable living. Sharing of knowledge among members and coordinated efforts in dealing with critical issues were sought in Thimpu.

However, despite such initiatives, progress in tackling climate change on a regional basis has not been satisfactory. The three-year SAARC Action Plan on Climate Change, adopted in 2008 and reiterated in 2011, which identified seven thematic areas of cooperation, putting special focus on adaptation, mitigation, and management of impacts and risks, has not been acted upon by member states. Though no summit took place after 2014, periodic meetings of External Affairs Ministers of the SAARC countries did take place. But unfortunately, the 37th SAARC Council of Ministers' Meeting in Pokhara (March 17, 2016) failed to reach any consensus about a target of reducing carbon emissions and a mutually agreed-upon plan of minimizing dependence on fossil fuels. Given the different levels of economic strength of these countries, a differentiated commitment could be worked out. But the SAARC countries failed to seize the moment and do something concrete at a time when the just-concluded Paris Climate Summit had generated worldwide hope about the resolve of those world leaders signing the Paris climate deal to do something positive to reduce carbon emissions and thus save future generations from the scourge of devastating climatic events.[6]

Thus, among the multitudes of policies adopted by SAARC, some have not been ratified, and many have not been made operational by the member countries. Because of lack of political cohesion, members have been hesitant to share technological know-how, and allegations have been made that India has not done enough to help its less resourceful neighbours. India, on its part, has complained of financial constraints and lack of cooperation. Imparting training to personnel of other countries and knowledge sharing are prominent areas where India, per estimation of experts, could do more. Critics also allege that SAARC summits were often paralysed because of hostile exchanges between India and Pakistan; hence, India, of late, has paid scant attention to SAARC and has tried to be more active in other regional bodies like the Association of South East Asian Nations (ASEAN) and the Bay of Bengal Initiative for Multi-sectoral Technical and Economic Cooperation (BIMSTEC) and has ignored SAARC. Consequently, a collective move by South Asian countries to cope with climate change has not received its due impetus.

India has defended its position strongly. The last SAARC summit, held in 2014 (Kathmandu), was attended by Indian Prime Minister Narendra Modi. The 2016 SAARC Summit was to be held in Islamabad. But in the wake of a dastardly terrorist attack on an Indian Army camp in Uri in Jammu and Kashmir on 18 September that year, in which 19 Indian soldiers died, India expressed its inability to participate in the summit due to 'prevailing circumstances'. India's objective was to escalate diplomatic pressure on Pakistan. Other South Asian countries, particularly Bangladesh, Bhutan, and Afghanistan, also expressed solidarity with India, and hence the Summit could not take place. India's position got support from various corners, and many international leaders endorsed its view that such vicious attacks on neighbours vitiate relations and leave no scope for regional cooperation.

The issue of organizing the SAARC summit came up for discussion during a meeting between Prime Minister Narendra Modi and his Nepalese counterpart K.P. Sharma Oli, who was in India on a three-day visit in April 2018. But as communicated to the press by India's foreign secretary Vijay Gokhale, Prime Minister Modi reiterated his sense of betrayal due to Pakistan's continued support of cross-border terrorism and made no commitments about a SAARC summit. Interestingly, India has recently held joint military exercises with the member countries of BIMSTEC, and since it brings together within its fold both South East Asian countries Thailand and Myanmar and four prominent South Asian countries, Bangladesh, Sri Lanka, Nepal, and Bhutan, apart from India, and significantly shuns Pakistan, analysts perceive it as India's diplomatic move of gradually abandoning SAARC, which cannot exclude Pakistan, and moving closer to another regional body which India considers to be more beneficial for its national interest as well as its regional aspirations (Gurjar, 2017). Climate change issues figure prominently in the agenda of BIMSTEC, and so, though the future of SAARC looks bleak at the present moment, observers believe that cooperation among South Asian countries on the climate change issue will continue within BIMSTEC and may actually receive a boost due to the congenial milieu of this new regional body. Experts of geopolitics also observe that India cannot and should not bypass SAARC since it incorporates all South Asian countries, and in the interest of the teeming millions inhabiting the region, climate change issues should be articulated through this body.

India's recent activism in the international arena

India was at the forefront of negotiations and deliberations among developed and developing countries which ultimately led to the Paris Agreement, formally known as the Conference of Parties (CoP) protocol on combating climate change, the world's first comprehensive regime on tackling the phenomenon within the United Nations Framework Convention on Climate Change. Adopted by 195 countries in Paris in December 2015, it is a watershed in the history of humanity's battle against climate change. The most significant objective of Paris Agreement is to keep global temperature increase well below 2 degrees Celsius and to pursue efforts to limit it to 1.5 degrees Celsius. The progress made towards the goal, per the pact, is to be reviewed every five years. As a significant part of the pact, the developed countries pledged $100 billion a year in climate finance for developing countries by 2020, with a commitment to further raise it in the future, something made possible by persistent lobbying by India along with other developing countries and their vociferous clamouring for climate justice. The prime minister of India has been at the forefront of negotiations and took the momentous decision of ratifying the treaty in 2016 on 2nd October, as a mark of respect to the father of the nation, Mahatma Gandhi, who was a strong advocate of leading lives in harmony with nature. Mr. Modi assured the world community that his country will go 'above and beyond' the 2015 Paris Accord to combat climate change and reduce its carbon footprint by 30–35 per cent by 2030. Over the last few years, India has taken several steps to tap non-fossil fuel sources and to increase its forest cover. The Ujjala Yojna has reduced burning of coal and has made the lives of rural women easier by giving them access to biogas and other renewable sources of energy. Scientific treatment of sewage and cleaning of rivers under Namami Gange projects has also contributed to saving the environment noticeably. The Green Skill Development Programme for skilling about 7 million youth in the environment, forestry, wildlife, and climate change sectors by 2021 is likely to create viable opportunities for skilled jobs and entrepreneurships in the environment sector and will make our youth the foot soldiers of the battle against climate change. India's big push to harness renewable sources of energy has found resonance among environmentalists and statesmen worldwide. In March 2018, leaders of several countries who receive abundant sunlight throughout the year congregated in New Delhi to mark the beginning of the International Solar Alliance, an innovative venture to harness the rich potential of solar energy on a grand scale. In October 2018, India's pioneering role in mitigating and adapting to climate change got international recognition when the United Nations celebrated Mr Modi with the Champions of the Earth Award, thereby acknowledging the spirited efforts of India's billions of ordinary citizens, environmental activists, government officials, and decision-makers to work together and respond to the challenge of climate change (Modi, 2018). In the recently held Conference of Parties 24, in the Polish city Katowice in December 2018, India emphatically proclaimed to the world that it was well on course to achieve all its climate targets much ahead of the deadlines it had set for itself. It also vehemently urged the developed countries to keep their commitments and fulfil their obligations to the world community in general and developing countries in particular, relating to providing financial and technological help to the developing countries so that India and other developing countries could keep their commitments and even go beyond that to saving our common home, the Earth.[7] Thus, India is acting as a true champion of the interests of the people of South Asia as well as those of the whole world.

Concluding observations

India has shown the grit and determination to be relentless in pursuing its climate agenda in the international arena, as it has put consistent pressure on United States under Trump, which has reversed its commitment to the Paris climate deal; other developed countries like Russia; and oil-exporting countries like Saudi Arabia and Kuwait, who are less than energetic about reaching the goal of carbon emissions just to ensure their own financial interests. India has secured for itself the role of a leader of the developing world and needs to lead by example by relentlessly pursuing its pro-environment agenda. As a matter of fact, India faces the challenge of accelerating the process of switching over to renewable sources of energy and taking many other suitable measures. It has already demonstrated to the world that the development requirements of an aspirant population can be and must be balanced with the larger global objective of saving the planet. Fortunately for India, neither its common man nor its politicians are in denial mode about the pervasive impact of climate change. Yet, an unholy nexus between some politicians, government officials, and developers of real estate still leads to land grabbing and destruction of water bodies and greenery that act as the lungs of our bustling cities. Authorities are sometimes compelled to take drastic measures, as happened in the city of Bangalore a few years back when the municipality undertook a demolition drive of buildings constructed illegally on encroached stormwater drains and lake beds across the city. Such constructions were blocking the flow of rain water, causing frequent inundation of this high-tech city. Our regulatory bodies need to be more vigilant to ensure that things don't escalate so as to warrant such measures. (Mukhopadhyay, 2017). People's activist role can also have a far-reaching impact on our struggle against climate change. In March 2019, children of many schools in India went on strike against climate change, demanding that the government take responsibility for the increase in carbon emissions. Present political dispensation has been castigated for steadily eroding coastal ecosystems via Coastal Regulation Zone (CRZ) notifications which permitted real estate and infrastructure works in the intertidal zone along India's coastline, despite warnings issued by scientists that rising sea levels could inundate settlements along the coast. The Land Acquisition and Rehabilitation and Resettlement Act has allegedly been misused to grab tribal land and hand it over to greedy real estate developers or energy conglomerates.[8] Citizens' alertness and activism against such nefarious moves are the needs of the day.

Interestingly, the climate change issue featured prominently for the first time in the electoral manifestos of both the Bharatiya Janata Party (BJP) and Indian National Congress, two major political parties of India before the 17th General Election in April–May 2019. The incumbent party, BJP, talked about 'Green Bonus', a long-pending demand. According to its manifesto, if the BJP came to power again, it would offer special help to Himalayan states in the form of a 'Green Bonus' to encourage forest conservation. The party promised to focus on the 102 most polluted Indian cities to resolve the issue of air pollution. After the resounding victory of the BJP, one expects the government of Prime Minister Narendra Modi to act on his promise.

India also needs to strike a healthy balance between its own national interests and those of its South Asian neighbours, to rise over petty squabbles, and to work in unison with them for a better future.

As a country boasting of its scientific and technological prowess, India must shoulder the responsibility of coordinating and invigorating the research conducted by other countries of South Asia under a unified command mechanism and ensure and facilitate implementation of well-thought-out measures for dealing with climate change. In view of the disturbing impact of internal migration as well as migration across countries within the region induced by climate change, and keeping in mind the material and human cost of such migrations, Indian political leaders must extend all possible help to the neighbouring states to mitigate the factors causing migration and support the rehabilitation projects undertaken by governments of these countries. India must take proper initiatives to bolster regional early warning system and disaster management mechanisms. It must lead by example through a proper modification of its energy consumption patterns, must incentivize investments and entrepreneurship in the area of renewable energy, and must resist the temptation to use technologies which are cheap but damaging for the environment. The paradigm of sustainable development needs to be followed by India so as to evoke a positive response and compliance from its less powerful neighbours. The responsible action of Indian leaders will enhance its regional as well as global stature of being a true eco-warrior. As India marches towards the goal of fulfilling the aspirations of its teeming millions, it must not forget its ancient wisdom of living in harmony with nature and must utilize its scientific and technological prowess with a sense of moderation and respect for Mother Earth so as to ensure sustainability and bequeath to our future generations a living planet. India gave the world the Vedas, the repertoire of profound knowledge about life, and Atharva-Veda's exhortation, 'The Earth is my Mother, her child am I', needs to be propagated and respected by India so as to synchronize material progress with the need to protect Mother Earth. The responsibility of averting the apocalypse and saving lives on this part of the planet rests on India.

Notes

1 IMF's World Economic Outlook Database, October 2019.
2 *The Hindu*, December 6, 2018.
3 Economic Survey 2017–18.
4 NAPCC, Ministry of Environment, Forest and Climate Change, Government of India.
5 Kathmandu Declaration, Official Website of SAARC.
6 *The Himalayan Times*, March 17, 2016.
7 *The Indian Express*, December 4, 2018.
8 EPW Engage, April 3, 2019.

References

Atteridge, A., Shrivastava, M. K., and Upadhyay, Pahuja N. 'Climate Policy in India: What Shapes International, National and State Policy?' US National Institute of Health Report, 2012. Available at: www.ncbi.nlm.nih.gov/pubmed/22314857 (Accessed on 23 January 2019).

Ghosh, Amitava. *The Great Derangement*. Gurgaon: Penguin Random House, India, 2016.

Government of India. *Economic Survey*. 2017–18. Available at https://ruralindiaonline.org/en/library/resource/economic-survey-2017-18-volume-ii (Accessed on 24 January 2019).

Government of India. 'National Action Plan on Climate Change (NAPCC).' Ministry of Environment, Forest and Climate Change. Available at: www.moef.nic.in/ccd-napcc (Accessed on 24 January 2019).

Gurjar, Sankalp. 'Is SAARC Doomed?' *The Diplomat*, April 1, 2017. Available at: https://thediplomat.com/2017/04/is-saarc-doomed/ (Accessed on 24 January 2019)

'Is India Ready to Tackle Climate Change?' *EPW Engage*, April 3, 2019. Available at: https://www.epw.in/engage/article/india-ready-tackle-climate-change (Accessed on 7 April 2019)

'Kathmandu Declaration.' Official Website of SAARC. Available at: http://saarc-sec.org/uploads/digital_library_document/03-Kathmandu-3rdSummit1987.pdf (Accessed on 24 January 2019).

Modi, Narendra. 'The Green State of Mind.' *The Indian Express*, October 4, 2018.

Mukhopadhyay, Jayita. 'The Greatest Challenge.' *The Statesman*, September 10, 2017.

Pandve, Harshal, T. 'India's National Action Plan on Climate Change.' Available at: https://www.ncbi.nlm.nih.gov/pmc/articles/PMC2822162 (Accessed on 24 January 2019).

Rangarajan, M. *Environmental Issues in India: A Reader*. New Delhi: Pearson-Longman, 2007.

The Himalayan Times. '37th SAARC Council of Ministers Meet Ends.' March 17, 2016.

The Hindu. 'India: The Third Largest Contributor to Carbon Emission.' December 6, 2018. Available at: http://mofapp.nic.in:8080/economic survey (Accessed on 23 January 2019).

Zafarullah, Habib, and Huque, Ahmed Shafiqul. 'Climate Change, Regulatory Policies and Regional Cooperation in South Asia.' *Public Administration and Policy,* 21(1) (2018): 22–35.

8 Evaluating the viability of shrimp aquaculture to impede climate change in the Sundarbans

Experiences from Bangladesh

Anindya Basu, Gouranga Nandy and Nabendu Sekhar Kar

Covering more than 10,000 km^2 of deltaic floodplains across Bangladesh and India, the Sundarban region – a UNESCO World Heritage Site, is the world's largest contiguous block of tidal halophytic mangrove forest. The largest estuarine delta of the world, formed from the rivers Ganga, Brahmaputra and Meghna, is intersected by a complex network of tidal waterways, creeks, canals and mudflats and also houses the Sundarban mangroves. The unique, dynamic ecosystem showcases a delicate balance between the freshwater flows of the tributaries and distributaries of Ganges-Brahmaputra riverine system and saline waters of the Bay of Bengal and is considered one of the most productive and important biosphere reserves of the world.

Ecologically, the mangrove forests are of utmost importance in this area of periodical tidal flooding, as they act as barriers to cyclones and tidal upsurges and also as a huge sink for absorbing carbon and other pollutants. Yet this unique and biodiversity-rich area is among the most threatened habitats, facing extensive mangrove exploitation due to biotic and abiotic pressure from the surrounding environment.

Aquaculture is believed to be a forerunner of 'blue revolution', as it supplies most of the world's marine products, helping to reduce poverty and food shortages in developing countries akin to what the 'green revolution' did to the agricultural sector in India. In the deltaic Sundarban region, too, aquaculture plays an important part as livelihood practice. The main objectives of this chapter are to evaluate the viability of aquaculture in impeding the socio-environmental crisis caused by climate change; to assesses in what way livelihood shift creates a greater rich-poor divide, further marginalizing ecological refugees; and to appraise how implementation of fishery regulation can balance the current situation.

Brief history of inhabitation

It is difficult to trace the true historical pattern of Sundarban inhabitation, as very little information prior to British colonial records is found. The earliest reference that could be traced back is to the epic Mahabharata (ca. 300 BC to 300 AD), where a reference to 'Gangasagar' (meaning the place where the river Ganges met the sea) was made (Dutt, 1989). Few attempts to convert the forest tracks of the Sundarban to cultivable plots were made during the Sultanate (c. 1206–1526) and Mughal (c.1526–1720) periods (Ascoli, 1921). It was the British who relentlessly pursued a policy of land reclamation through extensive deforestation and extension of cultivation in the Sundarban (Pargiter, 1934). As a result of this, human settlement in the Sundarban started to grow from the 18th century through clearing mangrove forests and constructing earthen embankments or dykes to protect the islands from tidal water intrusion. Embankments were made to facilitate rice production and prevent flooding in the croplands, while human habitation has greatly impacted the river-based irrigation and drainage systems (Ahmed, 2013). The recent mass settlement started in the early 19th century (Chaudhuri and Choudhury, 1994), and the settlers were mainly dependent on woodcutting, fishing and honey gathering as primary livelihood options.

The Sundarban is referred to as a 'climate change hotspot'. Elevation near the sea level makes this coastal tract highly vulnerable to climate-change impacts like sea level rise, coastal erosion, tidal surge and saline water inundation, posing threats to its fragile ecosystem and affecting human lives and livelihood. The inhabited islands of the delta are protected by man-made embankments against the ingression of saline water, which makes agriculture and aquaculture possible. The earthen embankments constructed back in the mid-19th century have already been worn out in several locations, and at the same time, the river beds have been raised through continuous siltation. In the Indian part of the Sundarban, out of a total of 3500 km of embankment surrounding the inhabited islands, 800 km are vulnerable to breach during high-intensity weather events like cyclones and storm surge (GoWB, 2019). Despite the ecological sensitivity of the region, the Sundarban is home to over 12 million people (of both India and Bangladesh) who depend on its natural resources for their livelihood, aquaculture being the most important one. Building resilience to climate change and related hazards which may crop up in future is the cornerstone of the strategy for enhancing the adaptive capacity of fish farmers.

Delineation of study area

The total area of the Sundarban (inclusive of both land and water) is approximately 10,200 sq. km; of which about 60 percent is in Bangladesh (around 6017 sq. km) and rest (around 4200 sq. km) in India (World Bank, 2014). To ensure the preservation of ecological distinctiveness of the Sundarban region, the Sundarban Reserve Forests (SRF) and Ecologically Critical Area (ECA) adjacent to the Sundarban have been demarcated in Bangladesh, and the Sundarban Biosphere Reserve (SBR) has been drawn up in West Bengal, India (Hussain, 2014); Fig 8.1.

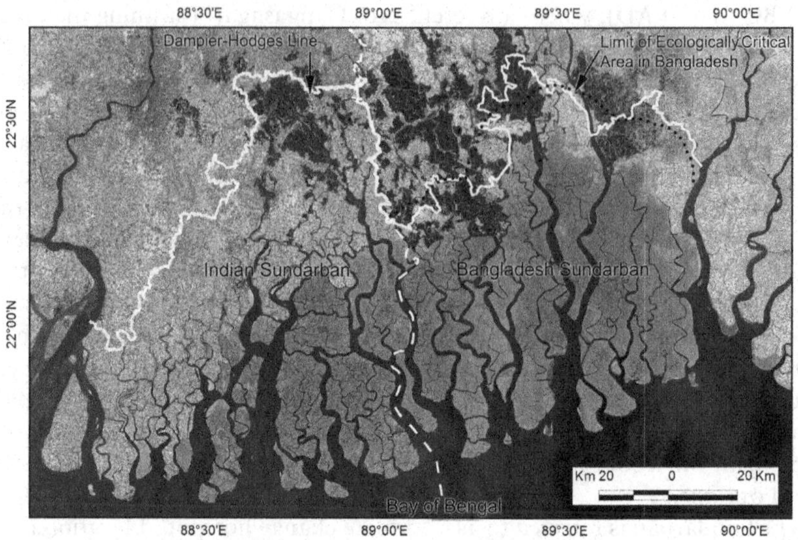

Figure 8.1 Sundarban Biosphere Reserve (SBR) of India consisting of both reclaimed and forest areas, demarcated by Dampier Hodges Line in the north. In Bangladesh, only the forest area known as Sundarban Reserve Forest (SRF) and a 10-km buffer area around the forest are demarcated as an ecologically critical area (ECA)

Sources: Authors. Landsat 8 OLI (p138-r044&p138-r045 of 08 March 2015 and p137-r044& p137-r045 of 17 March 2015)

The Indian part of the Sundarban is composed of 102 islands, of which 54 are inhabited. The Indian region is demarcated by the river Hooghly on the west, the Bay of Bengal on the south, the Ichamati-Kalindi-Raimangal rivers on the east and the Dampier-Hodges line on the north. It comprises 19 community development blocks, of which 6 are in North and 13 are in South 24-Parganas districts with a total of 190 Gram Panchayats and 1064 villages (GoWB, 2009; Danda et al., 2011). The Bangladesh part of the Sundarban comprises 20 upa-zilas (sub-districts) of five districts – Bagerhat, Satkhira, Khulna, Barguna and Pirojpur (BBS, 2011).

Role of climate change

The Sundarban is considered one of the worst climate change–affected areas of South Asia and is feared to be impacted by extreme climatic events like floods, droughts and cyclones (Kreft et al., 2015). The World Fish Centre identified four tropical Asian countries – Bangladesh, Cambodia, Pakistan and Yemen – as the most vulnerable economies depending on the impacts of climate change on fisheries (Allison et al., 2009). The climate of Sundarban is changing fast; the changes observed during the last two and a half decades reveal increasingly erratic weather, shortening winters, fast-rising trends of average daily minimum temperature, uncertainty of post-monsoon weather and rising incidences of partial break during mid-monsoon. The reports on climate change studies of the Sundarban confirm changes in air temperature, surface water temperature, rainfall and monsoonal patterns, freshwater availability, salinity regimes, cyclonic storms and depressions and sea level rise, along with erosion and accretion (Agarwala et al., 2003; IPCC, 2007b). Fish, as a poikilothermic animal, cannot regulate their body temperature through physiological processes and are thus directly influenced by the change of temperature. Increased temperatures and decreased levels of dissolved oxygen negatively affect pisciculture (Chowdhury et al., 2010). Climate change has an extensive influence on aquaculture, agriculture and fisheries sectors in developing countries that create pressure on their food security and livelihood (IPCC, 2007a). In aquaculture, high temperatures could also increase intensity and frequency of disease outbreaks of shrimp and adversely impact on water quality in source water bodies (Goggin and Lester, 1995).

Banes and boons influencing pisciculture

Tides: In the Sundarban, the tidal cycle is semi-diurnal with minor diurnal inequality. The Sundarban tides are also asymmetrical, with pronounced flood dominance. This means that the rising tide occupies a shorter time in a cycle, inducing faster landward velocity of the tidal current that turns the estuaries into sediment sinks. The tidal asymmetry also amplifies northward, suggesting an increasingly high rate of sedimentation in the upper part of the estuaries. This tidal pumping (Postma, 1967), that is, the net inflow of sediments, has profound implications on vertical accretion of the Sundarban region.

The regions that are reclaimed through embankments or polders prevent tidal inundation, and thus the natural process of delta building ceases. The areas remain lower than the highest high water and storm surge levels and are severely prone to saltwater ingression and flooding due to breaching or overtopping of the embankments. Though it cannot be denied that large-scale coastal polder construction in southwest Bangladesh since the 1960s has arrested tidal inundation, at the same time, it also resulted in loss of elevation. Auerbach et al. (2015) observed that there was a loss of 1–1.5 m of elevation for the reclaimed parts inside the polders compared to neighbouring mangroves, which continued to receive sediment through tidal inundation. This increases the region's vulnerability from rising sea levels and frequent storm surges.

Cyclones: Destructive cyclones hit the Sundarban area frequently, and the gusty winds generate huge waves which pile against the coast and finally terminate as storm surges, often inundating low-lying islands of the Sundarban (World Bank, 2014). The marginal embankments protect the reclaimed part of the Sundarban from cyclonic storm surges to certain extent, but when the landfall of a cyclonic storm coincides with the high tide period, the dykes tend to breach and inundate the areas behind them with the saline water. This was the case of tropical cyclone Aila (May, 2009), which caused extensive destruction and widespread inundation (IMD, 2010; Kar and Bandyopadhyay, 2015), especially in the Indian part.

Sea Level Rise: This has been an important aspect of the fallout of climate change. From the data available from tidal gauge stations and online portals (Nishat, 2019), a steady rise in relative mean sea level can be noted. The estuaries are extremely sensitive to changes in the mean sea level, as it has a great impact on tidal forcing and also influences sedimentation by setting up flood-dominated tidal asymmetry (Goodbred and Saito, 2012). This gradually rising sea level poses a serious threat of widespread future flooding.

Export-oriented shrimp farming was initiated from the mid-1980s in Bangladesh (Hossain et al., 2013), and the increasing sea level and resultant salinity increase initially boosted shrimp and other brackish water fish production. But if this salinity increase continues unabated, then it will negatively hamper shrimp production. Bhattacharyya et al. (2013) stated that the proliferation of shrimp farms in the reclaimed northern reaches of the Bhagna (upper Raimangal estuary) in western Sundarban has led to increased tidal discharge and accelerated erosion of the embankments along the channel.

Salinity: Though the Sundarban region is a coastal entity, the level of salinity from tidal influences is balanced by inflow of the upstream rivers and sufficient rainfall. Nishat (2019) has collated data from various sources and observed that the salinity of the Sundarban varies from 10–30 ppt depending on the season (heavy downpour in the monsoons decreases the salinity level, while reduced freshwater flow increases salinity during dry periods) and area.

There is a vicious circle around the salinity of the area. With increased salinity due to climate change (ill-distributed rainfall, reduced riverine flow, increased surface water salinity etc.), the resultant sea level rise affects the fish habitat, and on the other hand, increasing shrimp cultivation, which favours saline water, further increases soil salinity, making the region totally unsuitable for agriculture and even aquaculture. The recent study of Dasgupta and Shaw (2015)also corroborated the same findings; that is, linking fishery habitats and river salinity intrusion through the Sundarban estuaries has resulted in significant loss to fishery species.

Demographic profile of the region

In Sundarban, about 12 million people belong to the lowest socio-economic rung of society. The population density is extremely high in the Indian part at 1089 persons/sq km, while it is almost half, that is, 560 persons/sq km for the Bangladesh counterpart (Census of India, 2011; BBS, 2011 in Nishat, 2019).

Due to the fragile ecology and frequent climatic hazards, the livelihood opportunities of the residents are very limited. Moreover, they have to tread a very thin line of maintaining livelihood and not hampering the ecological balance through anthropogenic activities (Sen, 2016; Allan et al., 2013; Pathak and Kothari, 1998). In the Bangladesh part, people mostly work as fishermen, shrimp-fry collectors, wood-cutters and honey-gatherers (Islam, 2010), while their Indian counterparts are involved in fishing, crab collection, honey and beeswax collection and allied activities (Singh et al., 2010). The traditional fishing since the late 80s is being gradually replaced by the more profitable brackish-water shrimp farming (Bunting et al., 2011).

Locals perceiving the changes in climate

The locals who bear the direct brunt of climate change perceive it through their daily activities, as reported during field visits. The respondents described observed changes in patterns of seasonality in many ways, of which the most common changes experienced included warmer summers characterized by more intense heat, drier winters with decreasing rainfall and a shorter monsoon season with a decreasing trend in total rain in this season. The changes relating to the intensity and frequency of tropical cyclonic storms were perceived by them as severe storm surges often inundating the entire coastal area. Sea-level rise in their eyes means higher tidal water levels, often attributed to heavy siltation to river-beds, causing coastal erosion.

The shrimp farmers made certain interventions to tackle the risks associated with climatic change. To prevent fish from escaping ponds during periods of inundation caused by heavy rainfall or tidal flood, the farmers tend to increase the embankment height of the shrimp ponds or place nets around them. Due to intense heat during summer, the accelerated rate of evaporation increases the salt concentration in the shallow shrimp ponds, hampering production; to tackle this issue, deeper ponds are being dug so that the shrimp can take refuge (Shameem et al., 2015).

Initiation of fisheries

Fisheries and aquaculture are important contributors to global food supply, food security and livelihoods; low-income countries like India and Bangladesh are no exception. Aquaculture is one of the fastest-growing animal food-producing sectors; the annual growth averaged 6.1 percent for 2002–2012 (FAO, 2014) and above 8 percent for more recent years (HLPE, 2014). The growth is likely to continue (Merino et al., 2012) despite severe concerns about its continuing sustainability (Naylor et al., 2000). In Asia, the greatest producer of food fish by aquaculture is China (61 percent of total world market), with smaller contributions from India (6 percent), Vietnam and Indonesia (4.6 percent each) and Bangladesh (2.6 percent) (FAO, 2014). It is estimated that the global aquaculture sector provides around 27.7 and 56.7 million full- and part-time jobs (Philips et al., 2015). In Asia, a high proportion of production and sector employment is provided by aquaculture, which, apart from being an important source of food and nutrition, has also been able to generate foreign exchange through export markets. Fish products account for at least 15 percent of the animal protein consumed by more than 4 billion people, most in developing countries (FAO, 2012).

A source of significant growth in the global production of fish since the late 1980s has been culture fishery (aquaculture), with an average annual growth rate of 8.8 percent (FAO, 2014). The production from world food fish aquaculture more than doubled from 32.4 million tons in 2000 to 66.6 million tons in 2012, with an average annual growth rate of 6.2 percent during the period (FAO, 2014). In the Sundarbans, all kind of aquacultural practices can be seen, like inland and marine alongside freshwater, saline water and brackish water. An integrated approach is needed to reconcile increased conservation value and habitat connectivity, alongside livelihood betterment and poverty alleviation, involving a holistic understanding of the ecosystem and the dynamics between the ecological and closely related socio-cultural systems. The ecological and climatic conditions of coastal areas are extremely suitable for shrimp culture, which are essentially marine creatures, unlike prawns, with very low production cost (Islam and Wahab, 2005). Typical shrimp farms use a system of canals and sluice gates in the embankments to control the flow of water into and out of the *gher* (shallow shrimp pond).

Panning out of shrimp aquaculture

The entire stretch of the Sundarban is characterized by various kinds of aquacultural activities like coastal fisheries, brackish water aquaculture, estuarine and riverine fisheries, riverside prawn and shrimp seed collection, shrimp farming, crab harvesting and even freshwater aquaculture (Haque, 2003). Aquaculture and fisheries contribute 3.65 percent to national GDP and provide 60 percent of the animal protein consumed in Bangladesh (BER, 2016). The total farmed shrimp production in Bangladesh increased nine-fold over last 30 years (1986–2016) (Hossain and Hasan, 2017). Shrimp is the second largest export industry after textiles for Bangladesh (Rahman and Hossain, 2009), while marine products are ranked the 11th export-earning sector in India, with the shrimp sector contributing more than 69 percent of this, as reported by theMarine Export Development Authority (2017). As the international demand for shrimp is still going up steadily due to high demand from the European Union, United States and Japan, farmers are latching onto the opportunity and are also reaping a high return (GoI, 2014).

Initially, farmers practiced traditional methods of trapping shrimp post-larvae (PL) and fish seeds flowing in naturally to their farms with the tidal water. As they realized that switch to selective stocking would make shrimp farming more profitable, there was a sharp increase in the demand for wild PL through seed and brood stock collection. Currently, shrimp culture is heavily dependent on collection PL. Initially, wild PL were used, but that could not keep pace with rapidly expanding farms (Akhand and Hasan, 1992). The farmers believe that wild PL are more resistant to diseases and have good growth rates (Azad et al., 2009), and even to date, many South-east Asian countries depend on wild fry as their source of seed (World Bank, 2002). But many also believe wild PL harvesting is not ecologically viable (Primavera, 1998). Due to steady supply, the dependence on hatchery-produced PL has increased over time; about half of commercially viable tiger shrimp production is dependent on hatchery PL (Banks, 2003). The same is the case with shrimp feed; indigenously procured homemade feed mixtures of rice bran, wheat flour, oil cake and snail meat, due to price hikes and limited supply, has taken a backseat and commercial feeds gained ground (Naylor et al., 2000). Keeping in line with the policies of trade liberalization and export promotion, a generous supply of funds from national agents and international donors can be noticed; it is one of the main factors behind the extensive expansion of shrimp aquaculture (Pokrant and Reeves, 2003; Paul and Vogl, 2011). The other major factors that contributed to this growth area robust international market, high prices, cheap land and labour (Fleming, 2004; Islam, 2010), availability of shrimp fry (Islam and Wahab, 2005), favourable ecological and climatic conditions (Chowdhury et al., 2011) and formidable investments for upgrading production facilities (Alam and Pokrant, 2009).

Indian part

Fishery (fishing and aquaculture), both freshwater and brackish water, is treated as the backbone of the Indian Sundarban economy (Chand et al., 2012). The Indian Sundarban boasts around 172 species of fish, 20 species of prawn and 44 species of crab, including 2 commercial species. West Bengal is one of the front-running fish-producing states of India, with South and North 24 Parganas districts leading the pack (GoWB, 2014). Fishing activities are predominantly carried out in the reclaimed habitated areas; the main areas of traditional pisciculture are Sagar Island, Fraserganj, Bakkhali and Kalisthan, and major inland fish-landing regions in the Sundarban include Canning, Hariabhanga and Gosaba (World Bank, 2014). The Sundarban Biosphere Reserve is the top producer of fish and prawn, with both districts (South and North 24 Parganas) combined producing roughly 31 percent of the total inland fish and prawn production of West Bengal (Dubey et al., 2016). The estimated production of shrimp through aquaculture in West

Bengal increased from 12,500 metric tonnes during the year 1990–91 to 33,685 metric tonnes in 2009–10 (Chand et al., 2012).

In traditional brackish water shrimp cultivation, emphasis was on natural productivity. Thus, no supplementary feed was used, but as the yield was quite low, farmers started using supplementary feed and other inputs for higher production. Following the devastation of cyclone Aila (2009), trends of a shift in land use became more prominent; with the intrusion of saline water, the practice of saline aquaculture became more widespread. The species to which the highest emphasis is currently given is *Peneaus monodon*, or the black tiger shrimp (*bagda*), due to its high export demand and excellent unit value realization (Karim et al., 2013). Estuarine and riverside shrimp fry collection and prawn-seed collection has increased in the region over the last few years, driven by high profitability, but untenable practices have led to biodiversity loss and hampered other kinds of aquacultural practices in the Indian Sundarban (Cook, 2010). To safeguard the rights of vulnerable small-scale farmers, several regulations were passed; a licensing system for shrimp farms was initiated, but these have not been very effective (FAO, 2007).

Bangladesh part

Since the 1980s, shrimp farming has become an important aspect of livelihood options, as it helped in garnering high returns. Bangladesh ranks ninth in aquaculture production in the world (Pulhin and White, 2010). It is the fourth-largest inland fishery-producing nation in the world, and 2.7 percent of this fish is produced in the Sundarban. It has been estimated that about 55,000 tonnes were exported in 2011–12, fetching $454 million (Karim et al., 2013) and ranking as the second-highest export item after garments (BFRI, 2011). The inshore, estuarine and coastal fisheries of the Sundarban provide a major source of food and employment for a large chunk of the population. It is estimated that approximately 1.2 million people are directly dependent on shrimp production, with another 4.8 million household members supported by the industry (USAID, 2006). Banks (2003) stated that more than 0.7 million people were employed either directly or indirectly in shrimp farming and other related occupations. Even 10 years ago, nearly 1.2 million people were directly involved in shrimp production, with another 4.8 million household members supported by the industry (USAID, 2006). The present figure of employment in fishery and aquaculture activities is estimated at 14.7 million people (GoB, 2017), representing approximately 11 percent of the overall population of Bangladesh. As there is a ban on wood collection from the forest interiors, dependence on the fishery resources is quite high (Islam and Chuenpagdee, 2013); the poorer section of the fishing community depends on collection of crabs, molluscs and shells (CEGIS, 2015). While discussing worker composition, Pokrant and Reeves (2003) opined that the majority were involved in shrimp fry collection; the rest took part in processing, hatcheries or other related industries like pharmaceuticals, feed processing and ice manufacturing. Here also, like in the Indian counterpart, many are involved in collection of shrimp fry, as the yearly earnings are comparatively higher than the other avenues (Islam, 2010).

Shrimp cultivation was mainly practised in coastal areas, along the estuarine rivers and tidal creeks. A total of 123 polders were constructed in the 1970s by the Bangladesh Water Development Board (BWDB) under the Coastal Embankment Project (CEP). Approximately 5000 km of embankments were built, along with 1347 regulators, 1164 flushing inlets and 5937 km of drainage channels (Nuruzzaman, 2006). Polder construction was initially aimed at protecting low-lying farm plots from flooding and saline water intrusion. Gradually, the polders which provided protection from ingression of tidal waters gave way to the farming of *golda* or freshwater prawns (*Macrobrachium rosenbergii*). This variety is termed 'white gold' because of their high export value (Islam, 2008).

In several coastal districts, especially in the southwestern part – such as Khulna, Bagerhat and Shatkhira – shrimp farming has become the key economic activity (Afroz and Alam, 2013); Fig 8.2. After the initial modest

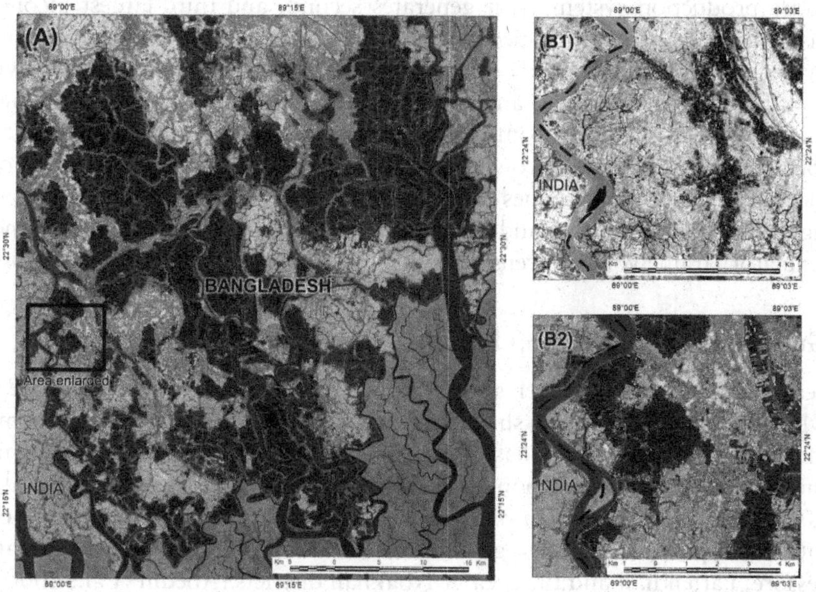

Figure 8.2 Proliferation of aquacultural firms along the south-western part of Bangladesh Sundarban forest margin, covering the districts of Satkhira, Khulna and Bagerhat (A). Change of land use between 1967 (B1) and 2015 (B2) in part of the Shyamnagar Upazilla of Satkhira District

Sources: Authors. Landsat 8 OLIp138-r045 of 08 March 2015(A & B2) Corona KH4A photograph, DS1038.2102DA-183 of 21-Jan-1967(B1)

beginning, shrimp farming rapidly gained popularity and by early 2000 reached its peak, covering about 200,000 ha of Satkhira, Khulna and Bagerhat districts in 2005 (Rahman and Hossain, 2009). It is estimated that 75 percent of exported shrimp are cultivated in these three districts (Islam et al., 2015). About the actual areal extent of shrimp farms, several reports have been published with varying data. However, a marked increasing trend is seen in all districts. In 1980, approximately 20,000 ha of land were being used for shrimp farming (Metcalfe, 2003). By 2010, this figure rose to 245,000 ha, representing a 1125% increase in area under cultured shrimp over the last 30 years (Belton et al., 2011). According to Karim et al. (2013), the area used for shrimp aquaculture in Bangladesh has expanded from about 20,000 ha in 1980 to nearly 244,000 ha in 2014, while in the report of Hossain and Hasan, (2017) it is stated that the land area under shrimp farming has increased from 70,331 ha in 1986 to 275,509 ha by 2016. The value of exports has also increased from $2.9 million in 1973 (EJF, 2004) to $380 million in 2005 (Ahmed, 2004). Although the total production was increasing, the rate of shrimp production was decreasing relative to the total area of farms, which is often attributed to climate change consequences (Rimi et al., 2013).

Hilsa fishing and dry fish (*shuntki*) production is an important part of the entire production system, as it generates second- and third-largest income in the fisheries segment, respectively, but that is also facing a decline due to overemphasis on shrimp fry collection (Hoq, 2003). Gradually, tiger prawn replaced both rice farming and freshwater prawn (*golda*) aquaculture, and this fast expansion, apart from causing environmental degradation, has led to social problems like severe conflict over land conversion, especially in areas of medium to low salinity zones (Maniruzzaman, 2012), and resultant human right violations (Rahman and Hossain, 2009). However, there has been no regulatory framework or licensing system put in place to rectify the situation.

Gradual change in shrimp farming system and farm size

Seasonal variation of water salinity is prevalent (Sohel and Ullah, 2012), which influences land uses; shrimp farming in rotation with paddies is common in the southwestern coastal part like that of Greater Khulna, while in the southeastern section, pond-based pisciculture is in vogue (Azad et al., 2009). In the case of *golda* or freshwater prawn aquaculture, the trend of integrated prawn-rice-fish-vegetables cycles is still on such as in Bagerhat, Jessore, Patuakhali and the greater Noakhali districts (Abedin et al., 2000). In Satkhira district, due to the presence of a high level of salinity, shrimp is grown year-round in the brackish water (Alam, 2002). Gradually, with commercial polyculture of shrimp, other finfish species or crustaceans are dying out (Azad et al., 2009).

Keeping parity with the evolution of pisciculture, there has been a change in the farm size and ownership patterns, too. From individually owned small farms and fish ponds (*beels*), larger fish farms (an amalgamation of smaller *ghers* or pond units converted from low-lying rice farms) on leased land under wealthy businessmen started to become more prominent (BCAS, 2001; Thomas et al., 2001; Alam and Phillips, 2004). In Bangladesh, 70 percent of shrimp farms use traditional and/or extensive culture techniques, 25 percent semi-intensive and 5 percent intensive (Hussain and Acharya, 1994). But over time, gradually, the farms are opting for more intensive techniques with greater input cost, as the yield rate is much higher. Increasing land values due to the profitability of shrimp have put rental prices out of reach for many of the area who do not have access to finance, thus transferring effective control of the lands to outsiders with the political connections and financial resources necessary to farm shrimp (Primavera, 1997).

Socio-economic consequences

Expansive shrimp aquaculture has several positive impacts on society, like increased income, improved food security and greater employment opportunities (Pokrant and Reeves, 2003; USAID, 2006; Belton et al., 2014), particularly opening up earning opportunities for women (Gammage et al., 2006; USAID, 2006).

Extensive shrimp production created enormous job opportunities for the locals both through shrimp seed production and post-harvesting processing, especially in southwestern Bangladesh (Keus et al., 2017; Ahmed and Flaherty, 2013; Bunting et al., 2017). An estimated 423,000 people are engaged in collecting shrimp fry, 40 percent of whom are women and children. They earn about $370, which is 16 percent higher than the typical yearly wage for day labour ($318) (USAID, 2008). On the opposite side, from a social perspective, the expansion of shrimp aquaculture has increased inequity (EJF, 2004; Murshed-e-Jahan et al., 2010), diminished land security (Hossain et al., 2013), reduced access to sharecropping opportunities (Samarakoon, 2004) and even intensified existing unequal gender and class relations (Datta, 2001; Belton et al., 2014).

The expansion of aquaculture came at a cost of marginalizing small and marginal farmers, gradually diminishing agricultural plots and creating a class of politically and financially strong shrimp farmers who hold larger tracts of plots. It was reported that more than one lakh farmers were displaced from the Satkhira district of Bangladesh since the initiation of shrimp farming (Islam and Bhuiyan, 2016). It has to be borne in mind that rice farming is more labour intensive than shrimp farming, so many were also rendered jobless. Halim (2004) opined that the group worst affected by the expansion of shrimp farming has been the landless as previously common land is taken for shrimp aquaculture (Halim, 2004). Physical displacement of local communities often led to violent disputes, causing major social and law and order problems (Deb, 1998; EJF, 2004).

The improvement in the economy came with a negative rider: the smaller farmers lost the rights over their plots as richer people from the city got into the business and took over the land (Islam, 2006); the independence of choosing a cropping pattern was taken from them. (Paul and Vogl, 2013).

Azad et al. (2009) noted that the Bangladesh government-owned (*khash jamin* – a common property resource) coastal lands were leased out to shrimp farmers who were mostly urbanites instead of being allocated to local landless people, as constituted in the Land Reform Board Act, 1989. So it was seen that larger aquacultural ponds were mostly owned by powerful sections like political leaders, relatives of bureaucrats and rich businessmen.

The rate of school dropouts increased formidably since the 1990s as the children got actively involved in shrimp seedling collection (Afroz and Alam, 2013), which made it difficult for future generations to opt for other livelihood opportunities. Due to this smash-and-grab attitude of wealthy businessmen, the local agriculturalists became 'white gold victims'.

A reduction of potable drinking water and farm land has been noted due to increased salinity and salt water intrusion. Conflicts between fisherman and shrimp fry collectors are also common, as the unscientific method of the latter significantly reduces the fisherman's catch. It has been estimated that 30 percent of all inland fish species, which provide 66 percent of the country's protein, are in danger of extinction (World Bank, 2006). Snails used for shrimp feed, due to high consumption, have nearly gone extinct in the region (Ahmed et al., 2008) (Fig 8.3).

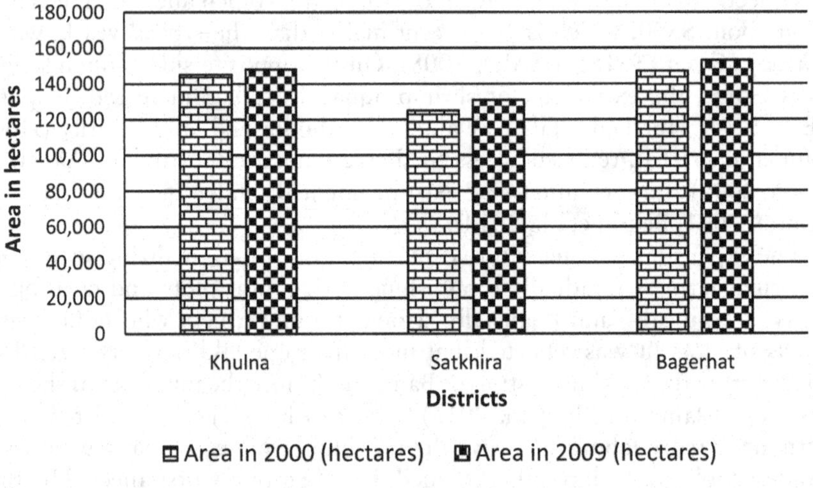

Figure 8.3 Gradual increase in areas affected by high salinity with extensive spread of shrimp firms between 2000 and 2009, along south-western part of Bangladesh covering districts of Khulna, Satkhira and Bagerhat

Source: Authors. Based on data from Mahmud, 2010

The euphoria of high income did not last for long, as shrimp disease broke out frequently, high salinity hindered crop rotation and there was significant loss of employment, too (Paul and Vogl, 2011). The return from livestock also dwindled as grazing became unsuitable in the saline areas. Efforts were made to develop integrated farming with fish-based aquaculture in *ghers* and paddy cultivation without retaining the saline water in the low-lying lands (Kabir et al., 2015; Ahmed and Flaherty, 2013; Ahmed et al., 2014), but the returns were not rewarding enough, as the land has become much too saline over the years. In the recent past, the government has been instrumental in addressing the issue of landlessness by supporting small farmers in their polyculture (Ahmed et al., 2014).

The Bangladesh coastal area was famous for black tiger shrimp (*Penaeus monodon*) production, which is a warm, brackish water species which grows well in salinity level 5–25 ppt and is highly vulnerable to diseases. As viable alternatives, Indian white shrimp (*P. indicus*) and western white shrimp (*P. vannamei*) came up. Western white shrimp for higher production, greater salinity resistance (0.5–45 ppt) and disease resistance capacity became extremely popular (Chanratchakool, 2003).

Hindrances around shrimp aquaculture

As shrimp aquaculture expanded, several studies have recognized wide-ranging environmental, economic and social impacts (Belton et al., 2014; Murshed-e-Jahan et al., 2014). Shrimp aquaculture has not really been able to deliver the expected outcome of alleviating poverty of the region. First, the productivity is quite low compared to the global context. Second, the shrimp industry is not only about the farmers: the chain contains the intermediaries who supply the fry and collect and distribute the produce, controlling the lion's share of the profit and marginalizing the actual producers. There are small (*faria*), medium (*aratdar*) and large aggregators in the shrimp value chain who usurp the profit share. Besides, shrimp production itself has created several environmental problems (Sen, 2010).

Environmental impacts have increasingly become a matter of concern for both the government and public with the fast expansion of coastal aquaculture dominated by shrimp farming (Paul and Vogl, 2011). The horizontal growth of shrimp farming is often unregulated, uncontrolled and uncoordinated (Deb, 1998; Metcalfe, 2003; Samarakoon, 2004). The ecological impacts are assumed to be exacerbated by climate change (Islam et al., 2014; Ahmed and Diana, 2015). The impacts of shrimp aquaculture in the Sundarbans region have created an 'ecological crisis' (Bala and Hossain, 2010), aggravating climate-related vulnerabilities of people living in the Sundarbans (Paprocki and Cons, 2014).

The major ecological issues involved are as follows.

Destruction of the mangrove ecosystem

The forest provides various essential commodities and services to communities like timber, firewood, pulpwood, thatching materials, medicinal herbs and so on. Also, the mangrove ecosystem, which is rich in biodiversity, serves as home to ecologically and economically viable species like birds, bees, crabs, crustacean, molluscs, turtles and others. Like other countries where aquaculture is practiced, the Sundarban also suffers from rampant loss of forest in general and especially mangroves for aquaculture (Hoq et al., 2001; Shahid and Islam, 2003; Hossain et al., 2013). Between 1980 and 2000, the FAO estimates that between 3.5 to 5 million ha were lost in Indonesia, Sri Lanka, India and Thailand; reducing mangrove cover by 28 percent in those four countries. Worldwide, 38 percent of mangroves are estimated to have been lost to shrimp farm conversion directly (EJF, 2006). There has been loss of traditional fuel sources in the Sundarbans due to shrimp farming, and thus now the local communities have no other option but to exploit the protected forest area (Thornton et al., 2009).

The Millennium Ecosystem Assessment (MEA, 2005) highlighted the vital role that ecosystem services play in society and how the present environmental degradation has affected the efficiency of 60 percent of the world's ecosystem services. Mangroves, which are widely recognized as extremely productive tidal wetland ecosystems, provide a variety of ecosystem services, including provisioning (Islam et al., 2018), like timber and fish production (Palacios and Cantera, 2017); regulatory, like wave attenuation, protection from cyclones, carbon sequestration and storage (Donato et al., 2011); and cultural services, like mangrove tourism and sacred forests. The mangrove area is extremely suitable for setting up fish nurseries (Abdullah et al., 2016), and reports suggest that the region enhances fishery yields, too, creating significant job opportunities (Mumby et al., 2004; Carrasquilla-Henao and Juanes, 2017). In spite of these factors, mangroves continue to be lost across the tropics (Hamilton and Casey, 2016) and are progressively threatened by conversion to aquaculture and agriculture, infrastructure development, overharvesting and anthropogenic climate change (Duke et al., 2014; Lovelock et al., 2015; Richards and Friess, 2016; Islam and Hossain, 2017; Thomas et al., 2017).

Land degradation due to increasing salinity

In some parts, shrimp farming happened by converting agricultural land into shrimp farms, resulting in the loss of agro-biodiversity and livestock (Azad et al., 2009; EJF, 2004). The increased shrimp farming in the polder areas led to widespread waterlogging, and the raised embankments of the region hinder the recession of water (IRIN, 2013). Many low-lying plots were turned into ponds, forcing residents to leave their homesteads; while some migrated completely, others adapted to the situation by switching

Table 8.1 Changing range of salinity in south-western Bangladesh covering districts of Khulna, Satkhira and Bagerhat

Salinity range (Ec:ds/m)	*Area in 2000 (hectare)*	*Area in 2009 (hectare)*
2–8	2,89,000	3,28,000
8–12	3,07,000	2,74,000
12–16	1,92,000	1,89,000
>16	87,000	1,01,000
	Σ6,14,900	Σ8,92,000

Source: Mahmud, 2010

their livelihood option to pisciculture. Usage of waterlogged land for commercial shrimp farming leads to increased water salinity, which causes soil to become highly saline and infertile for crop cultivation (Rahman et al., 2011; Kabir et al., 2016). Prolonged inundation inhibits the fixation of free nitrogen and halts mineralization, thus impairing soil fertility within a few years (Islam, 2003). The withdrawal of groundwater through pumping has lowered the groundwater table, and consequently fresh groundwater is contaminated by salt water (Flaherty et al., 2000). This has even led to the decline of grazing land and has substantially hampered livestock resources (Ali, 2006).

Major concerns in the Sundarbans were the trapping of salt within the farms for the entire season in order to continue with shrimp production and encroachment of public water bodies, canals and smaller rivers (Azad et al., 2009; Paul and Vogl, 2011). On the basis of salinity, the Bangladesh Sundarban is divided into three distinct ecological zones: a) Oligohaline zone (relatively freshwater), with salinity level less than 6250 micromhos, and the dominant species being Sundari (*Heritiera fomes*), which accounts for 60 percent of the total volume of commercial timber of the region. b) Mesohaline zone (moderately saline), which has a salinity level between 6250 and 12,500 micromhos, the leading species being Gewa (*Excoecaria agallocha*). c) Polyhaline zone (saline): here the salinity level exceeds 12,500 micromhos, and the leading species is Goran (*Ceriops decandra*). If the saline water moves inland, then Sundari, the oligohaline species, will be threatened, and if the salinity increases unabated, then the existing oligohaline zone might even be completely transformed into a mesohaline zone (Huq et al., 1999), and there would be gradual plant succession. The highly saline Indian part is chiefly dominated by Bain (*Avicennia sp.*) mangroves. Apart from affecting the productivity of the forest, the local people, dependent on forest products in the Sundarban, will also be exposed to economic uncertainties, as there would be a severe loss in merchantable wood from the Sundarban (Ahmed et al., 1998). This indiscriminate expansion of aquaculture is starting to become counterproductive (World Bank, 2014).

Acidification

Apart from the issue of increasing salinity of the water due to prolonged shrimp culture of the western white shrimp (*P. vannamei*) variety, increased acidity of soil is another issue. The exposure of the soil to excessive water leaches the alkaline element from it and retards crop growth, and high acidity leads to complete loss of productivity (Azad et al., 2009). To stabilize the pH and acid-base buffering capacity, a generous amount of lime is often added (Wilkinson, 2002). But farmers are ignorant about the regular application of phosphorous or other chemicals to improve pond water quality and control the occurrence of diseases.

Sedimentation and pollution

The runoff from upstream rivers, especially in the rainy season, brings in a formidable amount of sediment (Dewalt et al., 1996). Thus, when the water from the estuarine or riverine channels is stored in shrimp ponds or *ghers*, quick sedimentation at the bottom is noticed with the reducing water velocity (Dewalt et al., 1996). Sediments not only originate from river sources but also from the sludge accumulated on surrounding walls during each production cycle. These sludges accumulate due to various management practices, mainly in intensive farming systems, like high stocking density, feed application, aerator use, liming and so on (Funge-Smith and Briggs, 1998). As the usage of agrochemicals, antibiotics and disinfectants to treat water and prevent disease outbreaks increases, the amount of the suspended solids also rises, which in turn increases turbidity, reducing sunlight penetration, affecting the trophic structure and fauna and flora development (Dewalt et al., 1996; Wahab et al., 2003).

Effluents from shrimp ponds are often enriched in suspended solids: nutrients like ammonia, nitrate, nitrite (Páez-Osuna et al., 2003). If the effluents are directly discharged, then it can pollute the surrounding water and soil quality. The discharging effluents effectively reduce the dissolved oxygen in the water and create hyper-nutrification and eutrophication, which can harm the benthic species (Dewalt et al., 1996; Flaherty et al., 2000).

Due to lack of effective management, after a prolonged period of exploitation, the nutrients get exhausted, and the productivity of ponds declines. The remnants of the chemicals used accumulate in the water, which at times contaminates underground streams, reservoirs and even soil, giving rise to grave environmental consequences (Agarwal et al., 2019).

Loss of capture fishery stock

Collection of post larvae of wild-caught juvenile shrimp is preferred to hatchery-reared PL, since the former has a higher survival rate and is considered superior quality. Billions of shrimp fry are yearly collected from the wild, but indiscriminate collection through the usage of fine nets leads to loss of other fish species, affecting the overall structure of the trophic webs (Chowdhury et al., 2008). In this process, approximately 65–80 percent of the by-catch – these are other fish caught in the net during the search for post larvae – is ultimately completely discarded, causing huge losses to other aquatic species (EJF, 2004). Natural sources for shrimp collection are threatened by environmental pollution and overexploitation, causing a severe scarcity of wild seed supply (Hoq, 2007). Though there have been certain governmental regulations on this, the ban has mostly gone unenforced because hatcheries are only able to meet 50 percent of the demand.

Threat of genetic alteration

Due to shortage of local shrimp PL, farmers often depend on imported varieties. The introduction of foreign variations without proper precautions leads to gene pool interactions between native wild species and imported farm species, leading to gene pool alteration in the ecosystem through interbreeding (Deb, 1998). The native biodiversity is thus confronting environmental hazards due to the introduction of invasive species and modified genotypes (Naylor et al., 2005).

Disease outbreaks

Since the late 1990s, when a shift from an extensive nature of farming to an intensive one was noticed, shrimp farming has been plagued by outbreaks of various bacterial and viral diseases affecting both profitability and sustainability of shrimp farming operations (FAO, 1997; Mazid and Banu, 2002; GoB, 2007). Due to the scarcity of wild shrimp seeds and inadequate produce from the local hatcheries, many farmers are forced to depend on shrimp seed import from abroad. But often, this importing is carried out without quarantine, leading to the spread of viral and fungal diseases (Deb, 1998). In addition, the occurrence of diseases can be attributed to the stress generated by the degrading environmental quality along with unscientific shrimp pond management techniques (Páez-Osuna et al., 2003). The common diseases are red disease, white spot disease, soft shell, tail rot and black gill disease (Alam et al., 2007).

Direct from the grassroots

A survey revealed that governmental policies and non-governmental organization initiatives were not very successful in developing sustainable aquaculture practices in the Indian Sundarban (Dubey et al., 2016). Even the present field visits to Namkhana and Patharpatima blocks reflected similar results; in response to cross-sectional, multi-layered questionnaires, the locals complained about frequent cyclonic storm surges and resultant inundation, disease outbreaks, lack of exposure to technical aspects of aquaculture, increasing salinity due to unplanned monoculture and increasing incidences of social delinquency. From focus group discussion, the matter of conversion of the agricultural land of the marginal farmers to shrimp farming by the private sector came up, which has destabilized the agricultural economy, and very few alternative livelihood opportunities are available for income generation. So there is severe unemployment and underemployment in the region. This has led to lack of nourishment and health issues, distress migration, exploitative employment avenues and trafficking of women.

The farmers who are associated with shrimp culture do not have much exposure to institutional education, as due to poverty from early childhood, they get associated with piscicultural works. So they proclaimed that it is traditional knowledge derived mostly from experience that helps them in tiding over ecological and other crises. The farmers rue the absence of formal, institutional training, which they believe would have guided them to handle the polyculture of fish, shrimp and agriculture more efficiently. When questioned about the gradual shift from freshwater to brackish water aquaculture, most respondents expressed their awareness about the rising salinity related to such a shift but stated their helplessness, as the reducing land fertility was no longer suitable for cultivation, and for high yield and high monetary return, they had to carry out said practice. The women, who are engaged mainly as shrimp fry collectors to supplement their family income, due to the shortage of nursery ponds for fingerlings in the study area, admitted their vulnerability and complained that in this male-dominated field, they fail to command equal pay and have to toil hard for meagre amounts and often face sexual harassment from owners and middlemen, too. Beyond working in the shrimp farms directly, many women also get involved in net making. The children also, to support the family, take part in collection of prawn fry and are exposed to occupational hazards like waterborne diseases, skin infections and other contiguous diseases from a tender age (Das et al., 2016).

The respondents were univocal about the issue of middlemen. Due to lack of education and exposure, the farmers had little knowledge about global production scenarios, were not able to connect to the wide national and international export market directly and had to rely on intermediaries, often finding themselves at the receiving end of the vagaries of global economic conditions, which lowered their profit share drastically. Another issue which made their lives harder was the lack of efficient transportation facilities; as shrimp are highly perishable goods, faster transfer to the market is a necessity.

The residents of the southwestern districts of Bangladesh shared their experiences about how earthen coastal embankments, popularly known as polders, were constructed along the drainage channels and tidal estuaries to hinder free movement of the saline sea water and eventually increase the extent of agricultural plots. The elders shared how under the Bangladesh

Water Development Board, agricultural productivity increased by leaps and bounds. But then gradually, due to the non-functional sluice gates and other factors, the plots started to become waterlogged and cultivation took a back-seat. As the drainage flow got restricted and the entire area was flooded, the landowners had to look for alternative livelihood opportunities, and fisheries came into the foray. The direct interactions with the locals in Bangladesh were mainly made mainly in Satkhira, Bagerhat and Khulna districts.

Feedback about shrimp aquaculture was clearly varied; the landless locals and small landholders had negative views – under the condition of anonymity, they shared how they were tortured by the rich to forcefully opt for shrimp or even at times to sell their plots to them. They shuddered while describing how rampant hooliganism, atrocities and even heinous crimes like rape and murder have become common as the flow of cash has become concentrated in the hands of a few. They to some extent blamed themselves, as they also initially were tempted, and when they realized their mistake, it was too late to mend. The poorer section felt that the government also failed to protect them from the mayhem. The Khulna-Jessore Drainage Rehabilitation Project aka KJDRP (1994–2000), initiated for poverty reduction through increased agricultural production and creation of on-farm employment through reduction of man-induced salinity, met with limited success (Nandy, 2011).

Several movements were initiated in places like Paikgacha, Batiyaghata, Dahuri, Shaymnagar and Debhata against the rampant and forceful shrimp cultivation but could not sustain and reach the goal of limiting aquacultural practices. There have been several murders and extortion to initiate and continue shrimp aquaculture, the striking ones being Gobinda Dutta's killing at Jessore (1988), Karunamoyee Sardar's murder at Polder 22 (1990), the assassination of Jaber Sheikh in Khulna (1994) and the dual slaying of Kamrul Islam and Abdur Rashid in Bagerhat (1994), as pointed out by Nandy et al. (2007). The locals confided that they are aware of the dwindling land fertility and the profitability of shrimp farms. But they were in favour of polyculture, mixing agriculture with fisheries, and more importantly had grudges that they were denied of their share of profit by the richer sections that control the aquaculture value chain.

Ways ahead for sustainable shrimp aquaculture

Major problems faced by the shrimp aquaculture sector like landscape deterioration, waterlogging, salinization of land and water and degrading natural biodiversity have been identified. To tackle these issues and make shrimp aquaculture sustainable, a paradigm shift is needed away from current shrimp farming management practices (Hoq, 2007) to a more holistic and integrated approach which ensures environmental integrity and social cohesion. However, shrimp aquaculture has been relatively neglected in the development of sustainable practices (Philcox et al., 2010). Apart from improving the physical infrastructure of *ghers* or shrimp ponds, adoption of new management methods is also needed to improve and streamline the environmental performance of shrimp farming. Sustainable shrimp farming would be possible if technological improvements were carried out, adequate knowledge transfer were made through institutional framework and with appropriate monitoring of compliance with comprehensive policies and regulations.

Though there are several international regulations in place to curb the threats associated with shrimp farming, there is a lack of political will as well as proper legal sanctions to enforce them stringently. The noteworthy protocols are International Principles for Shrimp Farming, Standards for Responsible Shrimp Farming (FAO, 2007) and Code of Conduct for Responsible Fisheries (FAO, 1998).

As the sector involves both India and Bangladesh, a joint fisheries stock assessment can be done every few years to look at important aspects like quantity and quality of production, species diversity and dynamics, pollution level water quality and so on. This would in turn generate authentic data which will help to formulate cross-border management decisions, total production of fish and other aquatic organisms, fishery-wise production, year-wise production, species-wise production and fish population dynamics. The assessments need to generate basic information for management decisions.

Rampant mangrove destruction due to shrimp aquaculture has hampered the ecological balance of the entire region. Though shrimp farming is profitable, the avenues for alternative livelihood practices should be encouraged to maintain ecological poise. FAO has developed a Code of Conduct for Responsible Fisheries (CCRF) including aquaculture (FAO, 1995), calling for a total ban on shrimp farming in mangrove areas. To merge ecology with economics, the farms who complied with the norms became eligible to receive a Seal of Quality endorsement that considerably enhanced the appeal of their products to world markets with its international certification standards.

The rights and benefits of the local marginal shrimp farmers have to be safeguarded by freeing them from the noose of the middlemen. Infrastructural enhancements like hatchery development, ice industry development, improved transport facilities, storage facilities and product development (like grading, packaging, branding) along with effective value chain analysis will help in empowering the locals. Free institutional skill development training regarding scientific farming practices, exposure to marketing linkages and networks – local, national and international – and provision of insurance can help in transforming the sector for the better.

The poor shrimp farmers often depend on money lenders and middlemen; they are deprived of fair prices. As the items are highly perishable, the farmers, due to the absence of warehouses with proper ice backup and storage facilities, are bound to sell their catch to the money lenders. Facilities for micro-finance and access to easy credit with minimal interest will empower the locals to set up such cold chains and post-harvest establishments, curbing losses. If a few fish landing and marketing centres are established on the land side of the Sundarban through public-private partnership, the marketing chain will be more effective and rewarding for farmers.

It has been widely accepted that extensive saline aquaculture, which is more productive economically, increases the level of salinity, making the area ecologically unsustainable. To check the tendency of switching to the tiger shrimp variety, high taxes can be imposed alongside subsidizing fresh water varieties to a certain extent, encouraging a reverse shift (Sen, 2010).

Alternative and innovative culture systems need to be devised to make shrimp aquaculture production more sustainable, and the local farmers need to be sensitized to these. Prime importance has to be attached to fostering pond nurseries or backyard hatcheries to reduce the detrimental environmental impact of wild-caught PL. Newly invented floating cage culture can be tried on a pilot basis to check for viability. The development will depend on creating an investment climate to foster private-sector investment in state-of-the-art shrimp hatcheries (Knowler et al., 2009). The government has a major role in formulating appropriate policies and monitoring and enforcing them strictly, particularly on the issues of forceful land conversion and unscientific post-larvae harvesting.

Instead of depending on mechanical pumps and aerators, shrimp farmers rely on tidal exchange of water, natural entrainment of organisms with incoming tides, organic fertilizer and limited use of animal manure to enhance production. From nutrient budget studies, it can be seen that extensive shrimp ponds can act as a sink for nutrients (Wahab et al., 2003) and are environmentally beneficial where eutrophication due to anthropogenic activity is prevalent. As farmers use a minimal amount of antibiotics or chemicals in shrimp ponds in the Sundarban, it is often regarded as eco-friendly or organic farming. So there is a huge potential for certified organic shrimp farming development (Bunting et al., 2011).

Benessaiah and Sengupta (2014) adopted a social-ecological vulnerability framework to understand vulnerability and resilience consequences of deviations from common pool resources to aquaculture in Nicaragua. A similar approach can be taken to measure the actual viability of aquaculture to impede climate change in the Sundarban region. Human, financial and technological capital are undoubtedly important to modernize shrimp aquaculture in the Sundarban region, and it is praiseworthy to note how by mixing the species type and involving technological upgrades to a certain extent, the local farmers have been able to tackle the climate change menace quite effectively. The combined effort and mutual respect of all the stakeholders – farmers, exporters and government – will help the sector reach its full potential in the near future (Biao and Kaijin, 2007). It is undeniable that the climate is changing fast and the livelihood avenues of residents of low-lying ecologically sensitive regions have to adapt ways to confront that. This needs a holistic approach with a deeper understanding of the people's way of life and advocating for best practice recommendations (Abdullah, 2017). To tackle the human and ecological vulnerabilities in the critical hydroclimatic environment of the Sundarban, regulated shrimp aquaculture following scientific norms would be the best option.

References

Abdullah, Abu Nasar Mohammad, Stacey, Natasha, Garnett, Stephen T., and Myers, Bronwyn. 'Economic Dependence on Mangrove Forest Resources for Livelihoods in the Sundarbans, Bangladesh.' *Forest Policy and Economics,* 64 (2016): 15–24.

Abdullah, Abu Nasar, Myers, Bronwyn, Stacey, Natasha, Zander, Kerstin K., and Garnett, Stephen T. 'The Impact of the Expansion of Shrimp Aquaculture on Livelihoods in Coastal Bangladesh.' *Environment, Development and Sustainability,* 19(5) (2017): 2093–2114.

Abedin, J., Islam, S., Chandra, G., and Kabir, Q. E. 'Freshwater Prawn (*Macrobrachium rosenbergii*) Sub-Sector Study in Bangladesh.' *Greater Options for Local Development through Aquaculture (GOLDA).* Project Report, CARE, Bangladesh. Funded by the Department for International Development, UK, 2000.

Afroz, Tanzim, and Alam, Shawkat. 'Sustainable Shrimp Farming in Bangladesh: A Quest for An Integrated Coastal Zone Management.' *Ocean & Coastal Management,* 71 (2013): 275–283.

Agarwal, Nandini, Bonino, Chiara, Deligny, Ana, El Berr, Luisa, Festa, Charlotte, Ghislain, Manon, Homolova, Katarina, et al. 'Getting the Shrimp's Share. Mangrove Deforestation and Shrimp Consumption, Assessment and Alternatives.' In Y. Laurans and A. Rankovic (Eds.), *Sciences Po – Paris School of International Affairs – Biodiversity Values and Policies.* IDDRI and Sciences Po, Paris, 2019.

Agarwala, Shardul, Ota, Tomoko, Ahmed, Ahsan Uddin, Smith, Joel, and Van Aalst, Maarten. *Development and Climate Change in Bangladesh: Focus on Coastal Flooding and the Sundarbans.* Paris: OECD, 2003.

Ahmed, Ashfaque, Aziz, Abdul, Nowsher, Azm, Khan, Ali, Islam, Mohammad Nurul, Iqubal, Kazi, Md, and Islam, Shafiqul. 'Tree diversity as affected by salinity in the Sundarban Mangrove Forests, Bangladesh.' *Bangladesh Journal of Botany,* 40 (2011): 197–202.

Ahmed, Nesar. 'Freshwater Prawn Farming in Bangladesh: How Cultivation Is Financed.' *Shellfish News,* 18 (2004): 17–19.

Ahmed, Nesar. 'Linking Prawn and Shrimp Farming Towards a Green Economy in Bangladesh: Confronting Climate Change.' *Ocean & Coastal Management,* 75 (2013): 33–42.

Ahmed, Nesar, Demaine, Harvey, and Muir, James F. 'Freshwater Prawn Farming in Bangladesh: History, Present Status and Future Prospects.' *Aquaculture Research,* 39(8) (2008): 806–819.

Ahmed, Nesar, and Diana, James S. 'Threatening 'White Gold': Impacts of Climate Change on Shrimp Farming in Coastal Bangladesh.' *Ocean & Coastal Management,* 114 (2015): 42–52.

Ahmed, Nesar, and Flaherty, Mark S. 'Opportunities and Challenges for the Development of Prawn Farming with Fish and Rice in Southeast Bangladesh: Potential for Food Security and Economic Growth.' *Food Security,* 5(5) (2013): 637–649.

Ahmed, Nesar, Ward, James D., and Saint, Christopher P. 'Can Integrated Aquaculture-Agriculture (IAA) Produce "More Crop per Drop"?' *Food Security,* 6(6) (2014): 767–779.

Akhand, A. M., and Hasan, M. R. 'Status of Freshwater Prawn (*Macrobrachium* spp). Culture in Bangladesh.' In E. G. Silas (Ed.), *Freshwater Prawns* (pp. 32–41). Thrissur, India: Kerala Agricultural University, 1992.

Alam, S. M. Nazmul. 'Shrimp Based Farming Systems in South-Western Coastal Zone of Bangladesh.' Master Degree Thesis. Asian Institute of Technology, Thailand, 2002, 102p.

Alam, S. M. Nazmul, and Phillips, M. J. 'Coastal Shrimp Aquaculture Systems in Southwestern Bangladesh.' *Asian Fishery Science Journal,* 17 (2004): 175–189.

Alam, S. M. Nazmul, and Pokrant, Bob. 'Re-Organizing the Shrimp Supply Chain: Aftermath of the 1997 European Union Import Ban on the Bangladesh Shrimp.' *Aquaculture Economics & Management,* 13(1) (2009): 53–69.

Alam, S. M. Nazmul, Pokrant, Bob, Yakupitiyage, Amararatne, and Phillips, Michael J. 'Economic Returns of Disease-Affected Extensive Shrimp Farming in Southwest Bangladesh.' *Aquaculture International,* 15(5) (2007): 363–370.

Ali, Abu Muhammad Shajaat. 'Rice to Shrimp: Land Use/Land Cover Changes and Soil Degradation in Southwestern Bangladesh.' *Land Use Policy,* 23(4) (2006): 421–435.

Allan, A.A, Lim, M., Islam, K. M. N., and Huq, H. 'Livelihoods and Ecosystem Service Provision in the Southwest Coastal Zone of Bangladesh: An Analysis of Legal, Governance and Management Issues.' Working Paper – 1 ESPA Deltas, 2013.

Allison, Edward H., Perry, Allison L., Badjeck, Marie Caroline, Adger, W. Neil, Brown, Katrina, Conway, Declan, Halls, Ashley S., et al. 'Vulnerability of National Economies to the Impacts of Climate Change on Fisheries.' *Fish and Fisheries,* 10(2) (2009): 173–196.

Ascoli, F.D. *A Revenue History of the Sundarbans: From 1870 to 1920.* Bengal Secretariat Book Depot, Calcutta, 1921.

Auerbach, L. W., Goodbred Jr, S. L., Mondal, D. R., Wilson, C. A., Ahmed, K. R., Roy, K., Steckler, M. S., Small, C., Gilligan, J. M., and Ackerly, B. A. 'Flood Risk of Natural and Embanked Landscapes on the Ganges–Brahmaputra Tidal Delta Plain.' *Nature Climate Change,* 5(2) (2015): 153–157.

Azad, A. Kalam, Jensen, Kathe R., and Kwei Lin, C. 'Coastal Aquaculture Development in Bangladesh: Unsustainable and Sustainable Experiences.' *Environmental Management,* 44(4) (2009): 800–809.

Bala, B. K., and Hossain, M. A. 'Modeling of Food Security and Ecological Footprint of Coastal Zone of Bangladesh.' *Environment, Development and Sustainability,* 12(4) (2010): 511–529.

Banks, R. 'Brackish and Marine Water Aquaculture. Report on Fisheries Sector Review and Future Development.' Department of Fisheries, Dhaka, Bangladesh, 2003.

BBS (Bangladesh Bureau of Statistics. 'Population and Housing Census, Socio-Economic and Demographic Report National Series.' In *Statistics and Informatics Division (SID) Ministry of Planning*. Government of Bangladesh (Vol. 4), Dhaka, 2011.

BCAS (Bangladesh Centre for Advance Studies). 'The Cost and Benefits of Bagda Shrimp Farming in Bangladesh – An Economic, Financial and Livelihoods Assessment.' *Report Prepared for the Feasibility Study for the Shrimp Component of the Fourth Fisheries Project* (2001): 81.Belton, Ben, Ahmed, Nasib, and Murshed-E-Jahan, Khondker. 'Aquaculture, Employment, Poverty, Food Security and Well-Being in Bangladesh: A Comparative Study.' *Worldfish*. CGIAR Research Program on Aquatic Agricultural Systems. Penang, Malaysia. Program Report: AAS-2014-39(2014).

Belton, Ben, Karim, Manjurul, Thilsted, Shakuntala, Collis, W., and Phillips, M. 'Review of Aquaculture and Fish Consumption in Bangladesh.' *The World Fish Center*, (53) (2011): 71.

Benessaiah, Karina, and Sengupta, Raja. 'How is Shrimp Aquaculture Transforming Coastal Livelihoods and Lagoons in Estero Real, Nicaragua? The Need to Integrate Social–Ecological Research and Ecosystem-Based Approaches.' *Environmental Management,* 54(2) (2014): 162–179.

BER (Bangladesh Economic Review). 'Report of 2016.' Ministry of Finance, Government of Bangladesh. Available at: www.mof.gov.bd/en/index.php?option¼com_contentandview¼article and id¼367anditemid¼1 (Accessed on 30 December 2019).

BFRI (Bangladesh Fisheries Research Institute). 'Sustainable Management of Fisheries Resources of the Bay of Bengal.' BOBLME Project, Dhaka, Bangladesh, 2011.

Bhattacharyya, Somenath, Pethick, John, and Sarma, Kakoli Sen. 'Managerial Response to Sea Level Rise in the Tidal Estuaries of The Indian Sundarbans: A Geomorphological Approach.' *Water Policy,* 15(S1) (2013): 51–74.

Biao, Xie, and Kaijin, Yu. 'Shrimp Farming in China: Operating Characteristics, Environmental Impact and Perspectives.' *Ocean & Coastal Management,* 50(7) (2007): 538–550.

Bunting, Stuart W., Kundu, Nitai, and Ahmed, Nesar. *Rice-Shrimp Farming Eco Cultures in the Sundarbans of Bangladesh and West Bengal, India.* Colchester, UK: New Eco Cultures Case Study, Essex Sustainability Institute, University of Essex, Mimeo, 2011.

Bunting, Stuart W., Kundu, Nitai, and Ahmed, Nesar. 'Evaluating the Contribution of Diversified Shrimp-Rice Agroecosystems in Bangladesh and West Bengal, India to Social-Ecological Resilience.' *Ocean & Coastal Management,* 148 (2017): 63–74.

Carrasquilla, Henao, Mauricio, and Juanes, Francis. 'Mangroves Enhance Local Fisheries Catches: A Global Meta Analysis.' *Fish and Fisheries,* 18(1) (2017): 79–93.CEGIS. *Annual and Combined Monitoring Report.* Center for Environmental and Geographic Information Services, Bangladesh, 2015.

Census of India. 'District Census Handbook, 24 Parganas (North and South).' *Series-20 Part XII-B.* Directorate of Census Operations, West Bengal, 2011.

Chand, B. K., Trivedi, R. K., Dubey, S. K., and Beg, M. M. *Aquaculture in Changing Climate of Sundarban: Survey Report on Climate Change Vulnerabilities, Aquaculture Practices & Coping Measures in Sagar and Basanti Blocks of Indian Sundarban.* Kolkata, India: West Bengal University of Animal & Fishery Sciences, 2012.

Chanratchakool, Pornlerd. 'Advice on Aquatic Animal Health Care: Problems in Penaeus Monodon Culture in Low Salinity Areas.' *Aquaculture Asia,* 8(1) (2003): 54–56.

Chaudhuri, Amal Bhusan, and Choudhury, Amelesh. *Mangroves of the Sundarbans. Volume 1: India.* Bangkok: International Union for Conservation of Nature and Natural Resources (IUCN), 1994.

Chowdhury, Arabinda N., Mondal, Ranajit, Brahma, Arabinda, and Biswas, Mrinal K. 'Eco-Psychiatry and Environmental Conservation: Study from Sundarban Delta, India.' *Environmental Health Insights,* 2 (2008): 61–76.

Chowdhury, Md Arif, Khairun, Yahya, Salequzzaman, Md, and Rahman, Md Mizanur. 'Effect of Combined Shrimp and Rice Farming on Water and Soil Quality in Bangladesh.' *Aquaculture International,* 19(6) (2011): 1193–1206.

Chowdhury, M. T. H., Sukhan, Z. P., and Hannan, M. A. 'Climate Change and Its Impact on Fisheries Resource in Bangladesh.' In *Proceeding of International Conference on Environmental Aspects of Bangladesh (ICEAB10)* (pp. 95–98), Japan, 2010.

Cook, Jonathan, Cylke, Owen, Larson, Donald F., Nash, John D., and Stedman-Edwards, Pamela. *Vulnerable Places, Vulnerable People: Trade Liberalization, Rural Poverty and The Environment.* Washington, DC: The World Bank, 2010.

Danda, Anamitra Anurag, Sriskanthan, Gayathri, Ghosh, Asish, Bandyopadhyay, Jayanta, and Hazra, Sugata. 'Indian Sundarbans Delta: A Vision.' *World Wide Fund for Nature-India, New Delhi,* 40 (2011).

Das, Pritha, Das, Atin, and Roy, Souvanic. 'Shrimp Fry (Meen) Farmers of Sundarban Mangrove Forest (India): A Tale of Ecological Damage and Economic Hardship.' *International Journal of Agricultural and Food Research,* 5(2) (2016): 28–41.

Dasgupta, Rajarshi, and Shaw, Rajib. 'An Indicator Based Approach to Assess Coastal Communities' Resilience Against Climate Related Disasters in Indian Sundarbans.' *Journal of Coastal Conservation,* 19(1) (2015): 85–101.

Datta, Anjan. 'Who Benefits and at What Costs: Expanded Shrimp Culture in Bangladesh, in Grassroots Voice.' *Journal of Indigenous Knowledge and Development,* 3 (2001): 12–18.

Deb, Apurba Krishna. 'Fake Blue Revolution: Environmental and Socio-Economic Impacts of Shrimp Culture in the Coastal Areas of Bangladesh.' *Ocean & Coastal Management,* 41(1) (1998): 63–88.

Dewalt, Billie R., Vergne, Philippe, and Hardin, Mark. 'Shrimp Aquaculture Development and the Environment: People, Mangroves and Fisheries on the Gulf of Fonseca, Honduras.' *World Development,* 24(7) (1996): 1193–1208.

Donato, Daniel C., Boone Kauffman, J., Murdiyarso, Daniel, Kurnianto, Sofyan, Stidham, Melanie, and Kanninen, Markku. 'Mangroves Among the Most Carbon-Rich Forests in the Tropics.' *Nature Geoscience,* 4(5) (2011): 293–297.

Dubey, Sourabh Kumar, Chand, Bimal Kinkar, Trivedi, Raman Kumar, Mandal, Basudev, and Rout, Sangram Keshari. 'Evaluation on the Prevailing Aquaculture Practices in the Indian Sundarban Delta: An Insight Analysis.' *Journal of Food, Agriculture & Environment,* 14(2) (2016): 133–141.

Duke, Norman, Nagelkerken, Ivan, Agardy, Tundi, Wells, Sue, and Van Lavieren, Hanneke. *The Importance of Mangroves to People: A Call to Action.* Cambridge: United Nations Environment Programme World Conservation Monitoring Centre (UNEP-WCMC), 2014.

Dutt, K. *The Past of South 24-Parganas (In Bengali)* (Vol. 1, pp. 54–63). Baruipur, Kolkata: Sundarban Regional Museum, 1989.

Dutta, K. *Dakshin Chabbisparganar Ateet.* Sundarban Anchalik Shangrahashala, Vol. 1. Baruipur, 1989.

EJF (Environmental Justice Foundation). 'Desert in the Delta.' A Report on the Environmental, Human Rights and Social Impacts of Shrimp Production in Bangladesh, London, 2004.

EFJ (Environmental Justice Foundation). 'Mangroves: Nature's Defence Against Tsunamis.' A Report on the Impact of Mangrove Loss and Shrimp Farm Development on Coastal Defences, London, 2006.

FAO (Food and Agricultural Organization). *Code of Conduct for Responsible Fisheries*. FAO, Rome, 1995.

FAO (Food and Agricultural Organization). 'Disease Prevention and Health Management of Coastal Shrimp Culture.' *TCP/BGD/6714, Field Document*, Bangkok, 1997.

FAO (Food and Agricultural Organization). 'Towards Sustainable Shrimp Culture Development.' *Implementing the FAO Code of Conduct for Responsible Fisheries (CCRF)*, Rome, 1998.

FAO (Food and Agricultural Organization). 'International Principles for Responsible Shrimp Farming.' *Network of Aquaculture Centres in Asia-Pacific (NACA, UNEP, WB, WWF)*, Bangkok, 2007.

FAO (Food and Agricultural Organization). *The State of World Fisheries and Aquaculture*. FAO, Rome, 2012.

FAO (Food and Agricultural Organization). *The State of World Fisheries and Aquaculture: Opportunities and Challenges*. FAO, Rome, 2014.

Flaherty, Mark, Szuster, Brian, and Miller, Paul. 'Low Salinity Inland Shrimp Farming in Thailand.' *Ambio* (2000): 174–179.

Fleming, Claire. *Challenges Facing the Shrimp Industry in Bangladesh*. Dhaka, Bangladesh: American International School, 2004.

Funge-Smith, Simon J., and Briggs, Matthew R. P. 'Nutrient Budgets in Intensive Shrimp Ponds: Implications for Sustainability.' *Aquaculture,* 164(1–4) (1998): 117–133.

Gammage, Sarah, Swanburg, Ken, Khandkar, Mubina, Islam, M. Z., Zobair, Md, and Muzareba, Abureza M. 'A Gendered Analysis of the Shrimp Sector in Bangladesh.' *Greater Access to Trade and Expansion,* USAID, Dhaka, Bangladesh, 2006.

Goggin, C. L., and Lester, R. J. G. 'Perkinsus, A Protistan Parasite of Abalone in Australia: A Review.' *Marine and Freshwater Research,* 46(3) (1995): 639–646.

Goodbred, Steven L., and Saito, Yoshiki. 'Tide-Dominated Deltas.' *Principles of Tidal Sedimentology* (2012): 129–149. Dordrecht.

GoB (Government of Bangladesh). 'Annual Report (2005–2006).' Department of Fisheries, Ministry of Fisheries and Livestock, Dhaka, Bangladesh, 2007.

GoB (Government of Bangladesh). 'National Fish Week 2017 Compendium.' Department of Fisheries, Ministry of Fisheries and Livestock, Dhaka, Bangladesh, 2017.

GoI (Government of India). 'Handbook on Fisheries Statistics.' Ministry of Agriculture, Department of Animal Husbandry, Dairying and Fisheries, Krishi Bhavan, New Delhi, 2014.

GoWB (Government of West Bengal). 'Annual Report 2013–14.' Department of Fisheries, Aquaculture, Aquatic Resources and Fishing Harbours, South 24-Parganas, 2014.

GoWB (Government of West Bengal). 'District Disaster Management Plan – South 24 Parganas 2018.19.' District Disaster Management Department, Office of the District Magistrate, South 24-Parganas, 2019.

GoWB (Government of West Bengal). 'District Human Development Report South 24 Parganas.' Development and Planning Department, South 24-Parganas, 2009.

Halim, Sadeka. 'Marginalization or Empowerment? Women's Involvement in Shrimp Cultivation and Shrimp Processing Plants in Bangladesh.' In K. T. Hossain, M. H. Imam, and S. E. Habib (Eds.), *Women, Gender and Discrimination* (pp. 95–112). University of Rajshahi, Rajshahi, 2004.

Hamilton, Stuart E., and Casey, Daniel. 'Creation of A High Spatio-Temporal Resolution Global Database of Continuous Mangrove Forest Cover for the 21st Century (CGMFC-21).' *Global Ecology and Biogeography*, 25(6) (2016): 729–738.

Haque, M. E. 'How Fishers' Endeavors and Information Help in Managing the Fisheries Resources of the Sundarban Mangrove Forest of Bangladesh.' In N. Hggen, C. Brignall, and L. Wood (Eds.), *Putting Fishers' Knowledge to Work*. Fisheries Center Research Reports 11(1), (pp. 433–438). University of British Columbia, Vancouver, 2003.

High Level Panel of Experts on Food Security and Nutrition (HLPE). 'Sustainable Fisheries and Aquaculture for Food Security and Nutrition.' A Report by the High-Level Panel of Experts on Food Security and Nutrition of the Committee on World Food Security, Rome, 2014.

Hoq, M. Enamul. 'Sustainable Use of Mangrove Fisheries Resources of Sundarbans, Bangladesh.' *Tropical Agricultural Research and Extension,* 6 (2003): 113–121.

Hoq, M. Enamul. 'An Analysis of Fisheries Exploitation and Management Practices in Sundarbans Mangrove Ecosystem, Bangladesh.' *Ocean & Coastal Management,* 50(5–6) (2007): 411–427.

Hoq, M. Enamul, Nazrul Islam, M., Kamal, M., and Abdul Wahab, M. 'Abundance and Seasonal Distribution of *Penaeus monodon* Post larvae in the Sundarbans Mangrove, Bangladesh.' *Hydrobiologia,* 457(1–3) (2001): 97–104.

Hossain, M. A. R., and Hasan, M. R. 'An Assessment of Impacts from Shrimp Aquaculture in Bangladesh and Prospects for Improvement.' *FAO Fisheries and Aquaculture Technical Paper No. 618*, Rome, 2017.

Hossain, M. S., Uddin, M. J., and Fakhruddin, A. N. M. 'Impacts of Shrimp Farming on the Coastal Environment of Bangladesh and Approach for Management.' *Reviews in Environmental Science and Bio/Technology,* 12(3) (2013): 313–332.

Huq, S., Karim, Z., Asaduzzaman, M., and Mahtab, F. (Eds.). *Vulnerability and Adaptation to Climate Change for Bangladesh*. The Netherlands: Kluwer Academic Publishers, 1999.

Hussain, M. Z. 'Bangladesh Sundarban Delta Vision 2050: A First Step in Its Formulation – Document 2: A Compilation of Background Information.' In *IUCN, International Union for Conservation of Nature, Bangladesh Country Office* (pp. 1–192), Dhaka, Bangladesh, 2014.

Hussain, Z., and G. Acharya. *Mangroves of the Sundarbans. Volume two: Bangladesh*. IUCN Bangkok, 1994.

IMD (India Meteorological Department). *Annual Report, 2010*. New Delhi: Ministry of Earth Sciences, Government of India, 2010.

Intergovernmental Panel on Climate Change (IPCC). 'Summary for Policy Makers. In: Climate Change 2007. Mitigation.' In B. Metz, O. R. Davidson, P. R. Bosch, R. Dave, and L. A. Meyer (Eds.), *Contribution of Working Group III to the Forth Assessment Report*. Cambridge: Cambridge University Press, 2007a.

Intergovernmental Panel on Climate Change (IPCC). 'The Physical Science Basis.' In *Contribution of Working Group I to the Fourth Assessment Report 996*. Cambridge: Cambridge University Press, 2007b.

IRIN (Integrated Regional Information Networks). *Rethinking How to Help Water-Logged Communities in Bangladesh*. IRIN, Khulna, 2013.

Islam, G. M. Tarekul, Saiful Islam, A. K. M., Shopan, Ahsan Azhar, Rahman, Md Munsur, Lázár, Attila N., and Mukhopadhyay, Anirban. 'Implications of Agricultural Land Use Change to Ecosystem Services in the Ganges Delta.' *Journal of Environmental Management,* 161 (2015): 443–452.

Islam, K. M. N. *A Study of the Principal Marketed Value Chains Derived from the Sundarbans Reserved Forest*. Dhaka, Bangladesh: USAID, 2010.

Islam, M. Rafiqul. 'Managing Diverse Land Uses in Coastal Bangladesh: Institutional Approaches.' *Environment and Livelihoods in Tropical Coastal Zones* (2006): 237–248.

Islam, Md Shahidul. 'Perspectives of the Coastal and Marine Fisheries of the Bay of Bengal, Bangladesh.' *Ocean & Coastal Management,* 46(8) (2003): 763–796.

Islam, Md Saidul. 'In Search of 'White Gold': Environmental and Agrarian Changes in Rural Bangladesh.' *Society and Natural Resources,* 22(1) (2008): 66–78.

Islam, Md Shahidul, and Wahab, Md Abdul. 'A Review on the Present Status and Management of Mangrove Wetland Habitat Resources in Bangladesh with Emphasis on Mangrove Fisheries and Aquaculture.' *Aquatic Biodiversity II* (2005): 165–190. Dordrecht.

Islam, Mohammad Mahmudul, and Chuenpagdee, Ratana. 'Negotiating Risk and Poverty in Mangrove Fishing Communities of the Bangladesh Sundarbans.' *Maritime Studies,* 12(1) (2013): 7.

Islam, Mohammad Mahmudul, and Hossain, Mohammad Mosarof. 'Community Dependency on the Ecosystem Services from the Sundarbans Mangrove Wetland in Bangladesh.' In B. A. K. Prusty, R. Chandra, and P. A. Azeez (Eds.), *Wetland Science: Perspectives from South Asia* (pp. 301–316). New Delhi: Springer, 2017.

Islam, Mohammad Mahmudul, Sunny, Atiqur Rahman, Hossain, Mohammad Mosarof, and Friess, Daniel A. 'Drivers of Mangrove Ecosystem Service Change in the Sundarbans of Bangladesh.' *Singapore Journal of Tropical Geography,* 39(2) (2018): 244–265.

Islam, Mohammad Monirul, Sallu, Susannah, Hubacek, Klaus, and Paavola, Jouni. 'Vulnerability of Fishery-Based Livelihoods to the Impacts of Climate Variability and Change: Insights from Coastal Bangladesh.' *Regional Environmental Change,* 14(1) (2014): 281–294.

Islam, S.M. Didar-Ul, and Bhuiyan, Mohammad Amir Hossain. 'Impact Scenarios of Shrimp Farming in Coastal Region of Bangladesh: An Approach of an Ecological Model for Sustainable Management.' *Aquaculture International,* 24(4) (2016): 1163–1190.

Kabir, Kazi Ahmed, Saha, S. B., and Karim, M. 'Improvement of Brackish Water Aquatic Agricultural System Productivity and Environmental Adaptability.' In E. Humphreys, T. P. Tuong, M. C. Buisson, et al. (Eds.), *The Revitalizing the Ganges Coastal Zone: Turning Science into Policy and Practices. Conference Proceedings, CGIAR Challenge Program on Water and Food (CPWF)*, Colombo, 2015.

Kabir, Md Jahangir, Cramb, Rob, Alauddin, Mohammad, and Roth, Christian. 'Farming Adaptation to Environmental Change in Coastal Bangladesh: Shrimp

Culture Versus Crop Diversification.' *Environment, Development and Sustainability,* 18(4) (2016): 1195–1216.

Kar, Nabendu Sekhar, and Bandyopadhyay, Sunando. 'Tropical Storm Aila in Gosaba Block of Indian Sundarban: Remote Sensing Based Assessment of Impact and Recovery.' *Geographical Review of India,* 77(2015): 40–54.

Karim, M., Meisner, C. A., and Phillips, M. 'Shrimp (*Penaeus monodon*) Farming in the Coastal Areas of Bangladesh: Challenges and Prospects Towards Sustainable Development.' In C. A. Delaney (Ed.), *Shrimp Evolutionary History, Ecological Significance and Effects on Dietary Consumption* (pp. 57–88). New York: Nova Science Publishers, 2013.

Keus, E. H. J., Subasinghe, R., Aleem, N. A., Sarwer, R. H., Islam, M. M., and Hossain, M. Z. 'Aquaculture for Income and Nutrition: Final Report.' *Penang: World-Fish* (2017): 30.

Knowler, D., Philcox, N., Nathan, S., Delamare, W., Haider, W., and Gupta, K. 'Assessing Prospects for Shrimp Culture in the Indian Sundarbans: A Combined Simulation Modelling and Choice Experiment Approach.' *Marine Policy,* 33(4) (2009): 613–623.

Kreft, Sönke, Eckstein, David, Junghans, Lisa, Kerestan, Candice, and Hagen, Ursula. 'Global Climate Risk Index 2014.' *Who Suffers Most from Extreme Weather Events* (2015): 1Á31.

Lovelock, Catherine E., Cahoon, Donald R., Friess, Daniel A., Guntenspergen, Glenn R., Krauss, Ken W., Reef, Ruth, Rogers, Kerrylee, et al. 'The Vulnerability of Indo-Pacific Mangrove Forests to Sea-Level Rise.' *Nature,* 526(7574) (2015): 559–563.

Mahmud, I. 'Labanaktota Jorip-2009, Upakula Lona Panir Matra Pach Gun Baracha (Salinity Survey – 2009, Saline Water Range Increase 5 Times).' *The Daily Protom Alo*, Dhaka, March 22, 2010.

Maniruzzaman, M. 'Situation Analysis: Water Governance and Community Based Water Management.' CPWF-G3 Project Report, 2012.

Marine Export Development Authority. 'Annual Report 2017–18.' Ministry of Commerce and Industry, Government of India, 2017.

Mazid, M. A., and Banu, A. N. H. 'An Overview of the Social and Economic Impact and Management of Fish and Shrimp Disease in Bangladesh, with an Emphasis on Small-Scale Aquaculture.' In J.R. Arthur, M.J. Phillips, R.P. Subasinghe, M.B. Reantaso, and I.H. Macrae (Eds.), *Primary Aquatic Animal Health Care in Rural, Small-Scale, Aquaculture Development, FAO Fisheries Technical Paper 406* (2002): 21–25.

Merino, Gorka, Barange, Manuel, Blanchard, Julia L., Harle, James, Holmes, Rober, Allen, Icarus, Allison, Edward H., et al. 'Can Marine Fisheries and Aquaculture Meet Fish Demand from a Growing Human Population in A Changing Climate?' *Global Environmental Change,* 22(4) (2012): 795–806.

Metcalfe, Ian. 'Environmental Concerns for Bangladesh.' *South Asia: Journal of South Asian Studies,* 26(3) (2003): 423–438.

Millennium Ecosystem Assessment (MEA). 'Ecosystems and Human Well-Being: Current State and Trends.' In The *Millennium Ecosystem Assessment Series Vol. 5*. Washington, DC: Island Press, 2005.

Mumby, Peter J., Edwards, Alasdair J., Ernesto Arias-González, J., Lindeman, Kenyon C., Blackwell, Paul G., Gall, Angela, Gorczynska, Malgosia I.

'Mangroves Enhance the Biomass of Coral Reef Fish Communities in the Caribbean.' *Nature,* 427(6974) (2004): 533–536.
Murshed-e-Jahan, Khondker, Ben Belton, and K. Kuperan Viswanathan. 'Communication Strategies for Managing Coastal Fisheries Conflicts in Bangladesh.' *Ocean & Coastal Management,* 92 (2014): 65–73.
Nandy, Gouranga. *Jaaler fande: Paanisampad byabosthaponai antorjatik orthologⁿikari protisthaner Bhumika o khadyonirapotyae protikriya – KJDRP purbo o poroborti obostha porjalochona:* (*'Entrapment: Food Security Under IFI's Intervention In Water Resources Management – A Case of Pre and Post KJDRP.'*) IRV, Khulna, 2011.
Nandy, Gouranga, Ali, Shamsher and Farid, Tauhid Ibne. *Chingri o jono-orthoniti: Kar labh, kar khoti:* (*Shrimp And People Economy: Who Gains Who Losses.'*). Dhaka, Bangladesh: Actionaid, 2007.
Naylor, Rosamond, L., Goldburg, Rebecca J., Primavera, Jurgenne H., Kautsky, Nils, Beveridge, Malcolm C. M., Clay, Jason, Folke, Carl, Lubchenco, Jane, Mooney, Harold, and Troell, Max. 'Effect of Aquaculture on World Fish Supplies.' *Nature,* 405(6790) (2000): 1017–1024.
Naylor, Rosamond, Hindar, Kjetil, Fleming, Ian A., Goldburg, Rebecca, Williams, Susan, Volpe, John, Whoriskey, Fred, Eagle, Josh, Kelso, Dennis, and Mangel, Marc. 'Fugitive Salmon: Assessing the Risks of Escaped Fish from Net-Pen Aquaculture.' *Bioscience,* 55(5) (2005): 427–437.
Nishat, Bushra. *Landscape Narrative of the Sundarban: Towards Collaborative Management by Bangladesh and India.* Washington, DC: World Bank Group, 2019.
Nuruzzaman, M. 'Dynamics and Diversity of Shrimp Farming in Bangladesh: Technical Aspects.' In A. Atiq Rahman, A. H. G. Quddus, Bob Pokrant, and M. Liaquat Ali (Eds.), *Shrimp Farming and Industry: Sustainability, Trade and Livelihoods* (pp. 431–460). Bangladesh Centre for Advanced Studies (BCAS) and University Press Ltd., Dhaka, 2006.
Páez-Osuna, Federico, Gracia, Adolfo, Flores-Verdugo, Francisco, Lyle-Fritch, L, Alonso, Rosalba, Roque, A., and Ruiz-Fernández, Ana. 'Shrimp Aquaculture and the Environment in the Gulf of California Ecoregion.' *Marine Pollution Bulletin.* 46 (2003): 806–815.
Palacios, M. L. and Cantera, J. R. 'Mangrove Timber Use as an Ecosystem Service in the Colombian Pacific.' *Hydrobiologia,* 803 (2017): 345–358.
Paprocki, Kasia, and Cons, Jason. 'Life in a Shrimp Zone: Aqua-and Other Cultures of Bangladesh's Coastal Landscape.' *Journal of Peasant Studies,* 41(6) (2014): 1109–1130.
Pargiter, Frederick Eden. *A Revenue History of the Sundarbans, from 1765 to 1870. Supt., Govt. Print.* Calcutta: Bengal Govt. Press, 1934.
Pathak, Neema, and Kothari, Ashish. 'Sharing Benefits of Wildlife Conservation with Local Communities: Legal Implications.' *Economic and Political Weekly* (1998): 2603–2610.
Paul, Brojo Gopal, and Vogl, Christian Reinhard. 'Impacts of Shrimp Farming in Bangladesh: Challenges and Alternatives.' *Ocean & Coastal Management,* 54(3) (2011): 201–211.
Paul, Brojo Gopal, and Vogl, Christian Reinhard. 'Organic Shrimp Aquaculture for Sustainable Household Livelihoods in Bangladesh.' *Ocean & Coastal Management,* 71 (2013): 1–12.

Philcox, Neil, Knowler, Duncan, and Haider, Wolfgang. 'Eliciting Stakeholder Preferences: An Application of Qualitative and Quantitative Methods to Shrimp Aquaculture in the Indian Sundarbans.' *Ocean & Coastal Management,* 53(3) (2010): 123–134.

Philips, M., Henriksson, P.J.G., Tran, N., Chan, C.Y., Mohan, C.V., Rodriguez, U.-P., Suri, S., Hall, S., and Koeshendrajana, S. *Exploring Indonesian Aquaculture Futures.* Penang, Malaysia: Worldfish, 2015.

Pokrant, Bob, and Reeves, Peter. 'Work and Labour in the Bangladesh Brackish-Water Shrimp Export Sector.' *South Asia: Journal of South Asian Studies,* 26(3) (2003): 359–389.

Postma, Hendrik. 'Sediment Transport and Sedimentation in the Estuarine Environment.' *American Association of Advanced Sciences,* 83 (1967): 158–179.

Primavera, J. Honculada. 'Socio-Economic Impacts of Shrimp Culture.' *Aquaculture Research,* 28(10) (1997): 815–827.

Primavera, J. Honculada. 'Tropical Shrimp Farming and Its Sustainability.' In *Tropical Mariculture* (pp. 257–289). California, CA: Academic Press, 1998.

Pulhin, Juan and White, Patrick. 'Vulnerability and Adaptation of Aquaculture and Inland Fisheries to Climate Change in the Coastal Zone.' In *Impact of Global Change on Marine Resources and Uses, Brest.* 2010. Available at: https://www.academia.edu/12046344/Vulnerability_and_adaptation_of_aquaculture_and_inland_fisheries_to_climate_change_in_the_coastal_zone

Rahman, M. H., Lund, T., and Bryceson, I. 'Salinity Impacts on Agro-Biodiversity in Three Coastal, Rural Villages of Bangladesh.' *Ocean & Coastal Management,* 54(6) (2011): 455–468.

Rahman, M. M., and Hossain, M. M. 'Production and Export of Shrimp of Bangladesh: Problems and Prospects.' *Progressive Agriculture,* 20(1–2) (2009): 163–171.

Richards, Daniel R., and Friess, Daniel A. 'Rates and Drivers of Mangrove Deforestation in Southeast Asia, 2000–2012.' *Proceedings of the National Academy of Sciences,* 113(2) (2016): 344–349.

Rimi, Ruksana H., Farzana, Shazia, Sheikh, Md, Abedin, Md, and Bhowmick, Arjun Chandra. 'Climate Change Impacts on Shrimp Production at the South-West Coastal Region of Bangladesh.' *World Environment,* 3(3) (2013): 116–125.

Samarakoon, Jayampathy. 'Issues of Livelihood, Sustainable Development, and Governance: Bay of Bengal.' *AMBIO: A Journal of the Human Environment,* 33(1) (2004): 34–45.

Sen, Amrita. 'Exclusionary Conservation in the Sundarbans: Who Pays the Price?' *Economic & Political Weekly,* 51(53) (2016): 25–28.

Sen, Soham. 'Conservation of the Sundarbans in Bangladesh Through Sustainable Shrimp Aquaculture.' Policy Analysis Paper for Nishorgo Project, Bangladesh, 2010.

Shahid, Md A., and Islam, J. 'Impact of Denudation of Mangrove Forest due to Shrimp Farming on the Coastal Environment in Bangladesh.' In M. A. Wahab (Ed.), *Environmental and Socio-Economic Impacts of Shrimp Farming in Bangladesh* (pp. 67–75). Dhaka, Bangladesh: Bangladesh Centre for Advanced Studies (BCAS), 2003.

Shameem, Masud Iqbal Md, Momtaz, Salim, and Kiem, Anthony S. 'Local Perceptions of an Adaptation to Climate Variability and Change: The Case of Shrimp Farming Communities in the Coastal Region of Bangladesh.' *Climatic Change,* 133(2) (2015): 253–266.

Singh, Anshu, Bhattacharya, Prodyut, Vyas, Pradeep, and Roy, Sarvashish. 'Contribution of NTFPS in the Livelihood of Mangrove Forest Dwellers of Sundarban.' *Journal of Human Ecology,* 29(3) (2010): 191–200.

Sohel, Md Shawkat Islam, and Ullah, Md Hadayet. 'Ecohydrology: A Framework for Overcoming the Environmental Impacts of Shrimp Aquaculture on the Coastal Zone of Bangladesh.' *Ocean & Coastal Management,* 63 (2012): 67–78.

Thomas, M. A., Macfadyen, G., and Chowdhury, S. 'The Costs and Benefits of Bagda Shrimp Farming in Bangladesh.' *Bangladesh Centre for Advanced Studies, Fourth Fisheries Project, Dhaka, Bangladesh* (2001): 82.

Thomas, Nathan, Lucas, Richard, Bunting, Peter, Hardy, Andrew, Rosenqvist, Ake, and Simard, Marc. 'Distribution and Drivers of Global Mangrove Forest Change, 1996–2010.' *PLOS ONE,* 12(6) (2017): E0179302.

Thornton, C., Shanahan, M., and Williams, J. *From Wetlands to Wastelands: Impacts of Shrimp Farming.* London: SWS Bulletin, Environmental Justice Foundation, 2009.

US Agency for International Development (USAID). 'A Pro-Poor Analysis of the Shrimp Sector in Bangladesh.' Greater Access to Trade Expansion, Development & Training Services, Arlington, USA, 2006.

US Agency For International Development (USAID). 'Summary Report -2008.' Work Planning Session with Department of Environment on Co-Management of Protected Areas, Dhaka-IPAC, 2008.

Wahab, Md Abdul, Bergheim, Asbjorn, and Braaten, Bjorn. 'Water Quality and Partial Mass Budget in Extensive Shrimp Ponds in Bangladesh.' *Aquaculture,* 218(1–4) (2003): 413–423.

Wilkinson, S. 'Aquaculture Fundamentals: The Use of Lime, Gypsum, Alum and Potassium Permanganate in Water Quality Management.' *Aquaculture Asia,* 7(2) (2002): 12–14.

World Bank. 'To Analyze and Share Experiences on the Better Management of Shrimp Aquaculture in Coastal Areas.' Shrimp Farming and the Environment – A World Bank, NACA, WWF and FAO Consortium Program Synthesis Report, 2002.

World Bank. 'Bangladesh: Country Environmental Analysis.' *Bangladesh Development Series: Paper 12,* Dhaka, 2006.

World Bank. Report 88061 – 'Building Resilience for Sustainable Development of the Sundarbans through Estuary Management, Poverty Reduction, and Biodiversity Conservation' – Strategy Report, South Asia Region, Sustainable Development Department Environment & Water Resources Management Unit, Washington, DC, 2014.

9 Sri Lanka's natural vulnerabilities

A political case study of the 2004 tsunami, its aftermath and responses

Sreya Maitra

There may be various vantage points to launch oneself into a narrative about an episode that remains ensconced in history, public memory and official records. The simplest one, which is recounting the chronology of events, would seem important but incomplete, unless we can contextualise its significance and analyse the larger implications in retrospect. The 2004 tsunami that gripped Sri Lanka and Aceh in Indonesia poses a very similar challenge. The problems of human security and environmental damage that it posed were harsh realities which the people and the governments had to grapple with; the tsunami cared neither about academic theoretical paradigms nor geographical boundaries. However, the post-tsunami intellectual pursuits and scholastic theorisations have constantly searched for classifications and recognisable patterns for ease of refined understandings and creating a meaningful body of literature. (See Hyndman, 2007; Tudor Silva, 2009; Khazai et al., 2006; Kuhn, 2009; Stirrat, 2006.) In the world of political science and international relations, the tsunami of 2004 in itself has not found extensive research. However, the expanding literature on non-traditional or non-conventional security threats (within the larger debate on the broadening of the security agenda) and the post-modern ideological movements on environmentalism have discussed the grave consequences of natural disasters like tsunamis in different essays. This is particularly relevant for South Asia, as it is endowed with a complicated security environment (Vinod, 2009: 143). The first section of the chapter takes this standpoint to theoretically contextualise the tsunami as one of the many environmental hazards which threaten humankind, along with climate change. The actual dynamics of the tsunami that hit Sri Lanka in December 2004 form the second section. In the third section, the chapter dwells on the political milieu in Sri Lanka at the time of the tsunami. Here, there is a brief discussion on the literature which exists on the inter-relationship between political conflict and environmental disasters. This corroborates the analysis of the ground-level impact of the tsunami on the then-ongoing Sinhala-Tamil war in Sri Lanka. The fourth section explores the international responses to the tsunami and the politics of aid and rehabilitation. The final section tries to assess the lessons learnt in disaster management and levels of awareness regarding climate change at the policy level in South Asia in general and more specifically in Sri Lanka. Thus, the aim of the chapter is to provide a detailed study of the disaster of 2004 and highlight the environmental hazards still plaguing Sri Lanka. It emphasises the urgent need to articulate ways and means of slowing down environmental degradation and resultant climate change. A refined and nuanced understanding of techniques of disaster management is needed, along with comprehensive, collaborative, action-oriented responses on the part of all stakeholders in South Asia to cope with climate change. The study argues that a pro-active role has to be consistently pursued at two key levels: disaster prevention and disaster management. While for disaster prevention, the role of science and technology, seismology and weather pattern predictions would be crucial, at the level of disaster management, the target is to make local governments sufficiently empowered and equipped for disaster-related responses. There must also be effective motivation and ground-level mobilisation of the civil society to implement management techniques and timely prevent the escalation of the crisis.

Theoretically approaching environmental problems

As already mentioned, the environment has found space in the academic discourses of international relations (IR) in environmental movements and non-traditional security. These may be briefly discussed at this juncture.

The rapid pace of industrialisation, economic development, urbanisation and mechanisation which gripped the First World and by default engulfed its colonies (constituting the Third World) since the seventeenth century reached its climax with the post-Second World War reconstruction and the undercurrents of race for nuclear weaponisation during the Cold War. The dangers of pollution, dwindling reserves of fossil fuels, deforestation and animal extinction were increasingly visible through natural aberrations like global warming, ozone depletion and acid rain (Heywood, 2002a; 2002b). There were two subsequent developments as a response to this, which was theorised in two related strands of literature. Environmentalism (1960s onwards) expressed itself through the ecological or Green movement, eco-feminism, eco-conservatism and eco-socialism. It developed an eco-centric worldview that portrayed the human species as a part of nature, and nature itself was not just a convenient resource available to satisfy human needs (Heywood, 2002b: 63). It reiterated that environmental concerns have to be acute, because economic growth is endangering both the survival of the human race and the very planet it lives on. A new generation of activist pressure groups also developed, loosely called 'eco-warrior' groups, who collaborated with larger environmental groups like Greenpeace, Friends of Earth and the Worldwide Fund for Nature to give momentum to the green movement. From the 1980s onwards, environmental questions have been kept high on the political agenda by green parties, which now exist in most industrialised countries, often modelling themselves on the pioneering efforts of the German Greens (Heywood, 2002a: 253).

The environment as an agenda for security arose in the 1980s as environmental disasters had severe impact on human lives, their physical security in terms of habitat and survival and psychological security in terms of the anxiety of coping with losses and being exposed to similar disasters in the future. Moreover, as the frequency and intensity of environmental disasters became more acute, government responses had to be prompt and holistic. Thus, 'securitising environment' was enthusiastically discussed among scholars of security. The Copenhagen school of security made a prominent analytical contribution in this regard. Ole Waever had put forward the concept of securitisation (and desecuritisation) in an essay in 1995, and this was conceptually fused with Buzan's sectoral analysis of security in their 1998 work entitled *Security: A New Framework of Analysis*. The central concern of the book pertains to the dilemma that while the broadening of security is necessary, there are 'intellectual and political dangers in simply tacking the word security onto an ever wider range of issues'(Buzan et al., 1998: 1). In exploring the issues that may be amenable to securitisation, the book delves into a sectoral analysis of security in keeping with the broadening agenda. The authors analyse five sectors: military, environments, societal, economic and political, and try to locate the security agenda in each. In the environmental sector, the authors concede that securitisation of environmental values has a very short history compared to the other four sectors. And, given fluctuations in environmental threat perceptions, the field and discourse are developing. Scientific and political agendas overlap in the environmental sector, and it is the former that underpins the securitising moves, as it is here that the questions of evidence and proof are provided. The referent object (what is securitised) is the environment itself, but in many cases, the nexus between civilisation and environment is also sought to be securitised. Recognising that environment as an issue area evokes contradictory responses among actors, Buzan et al. identify 'lead actors' like states or global, environmental, epistemic community and non-governmental organisations and activists which promote effective international action on an environmental issue in specific cases. On the other hand, there are veto actors and veto coalitions comprising states and firms which play down threats posed by environmental issues (Buzan et al., 1998: 77–78). The logic of vulnerability to the environment hinges on threats from human activity to the natural systems of structures of the planet which bring about changes which do seem to pose existential threat to (parts of) civilisation. This accords with the need to talk about environmental security and establishes a circular relationship of threat between civilisation and the environment in which the process of civilisation involves a manipulation of the rest of nature; in many cases, this has achieved self-defeating proportions (Buzan et al., 1998: 80). Successful securitisation has occurred mainly at the local level in this sector.

But securitising moves are mostly at the global level (Buzan et al., 1998: 90). The cognitive dimension that feeds regional environmental regimes is global, but the sizes of the security complexes are determined bottom up. The environment modified by human interference sets the conditions for socio-political life. In the absence of environmental security, there will be conflicts over threats to non-environmental existential values (Buzan et al., 1998: 77–78).

The observations of Buzan et al. can be evidenced in the divergences between the efforts of the international and local communities. While the former has tried to create a global awareness regarding environmental risks and climate change and intervene in post-disaster situations through aid and humanitarian action, immediate post-disaster management has largely happened at the levels of the stakeholders in the affected provinces and the state. The politics regarding environmental security and human lives thus play out mainly at these levels. However, this is neither necessary nor desirable. The events revolving around the handling and aftermath of Sri Lanka's tsunami of 2004 serve as a case in point.

Tsunami engulfing Sri Lanka: December 2004

Tsunami is a Japanese term that denotes a series of waves generated by underwater seismic disturbances. In case of the tsunami of December 2004, there was an interface of the tectonic plates between India and Burma. Seismologists with the United States Geological Survey said the ocean west of Sumatra and the island chains to its north was a hot zone for earthquakes because of a nonstop collision occurring there between the India plate, beneath the Indian Ocean seabed, and the Burma plate, under the islands and that part of the continent. The India plate is moving at about two inches a year to the northeast, creating pressure that releases, sporadically, in seismic activity (Waldman, 2004). But the Asian tsunami that struck the coasts of Sri Lanka in December 2004 was an especially devastating earthquake, the fourth most powerful in 100 years. It was the worst natural disaster to have hit Sri Lanka for centuries. Before this, occasional floods had troubled the island country, but tsunamis, earthquakes and volcanoes had been at bay. But as geographical observations about environmental disasters narrate, island countries are most vulnerable and face the danger of disappearing from the world map (Chatterjee, 2010: 272). One-third of the coastline was severely affected by the huge waves. The tsunami hit five other countries across Asia: Indonesia, India, Thailand, Malaysia and the Maldives. Tourists, fishermen, hotels, homes and cars were swept away by walls of water unleashed by the 8.9-magnitude earthquake, centred off the west coast of the Indonesian island of Sumatra. Sri Lanka was within 1,000 miles west of the epicentre and thus one of the worst hit. The government declared it a national disaster (*The Guardian*, December 26, 2004). In retrospect, it has been recorded that the tsunami killed 35,000 people and left

900,000 homeless. This is quite a lot in a nation that has only 19 million people, with the majority of them living near the coast. Nearly 40 percent of the dead were children (Hays, 2008). According to a World Bank report, approximately 1 million out of a national population of 19 million people (5.3%) have seen their lives affected by the tsunami in some way (World Bank, 2005; INUC, 2005, cited in World Bank, 2005: S829–S844). The tsunami completely destroyed 80,000 housing units and damaged an additional 40,000, as well as schools, private-sector buildings and public-sector buildings (Information Paper No.3, 2005, cited in World Bank, 2005: S829–S844).

The tsunami struck Sri Lanka about 1 hour and 45 minutes after the Sumatran earthquake that created it. It hit the east coast first at around 8:45 am local time. It took another 30 minutes for it to whiplash around the island to the south coast, which was struck around 9:15 am local time. The shock waves from the earthquake were enough to lift the earth four inches in Sri Lanka. The waves travelled up to 800 kph in deep water and were 300 kilometres wide. The largest waves were 6 meters. They surged inland up to a kilometre. Unlike Thailand, where the waves crashed ashore like large Hawaiian waves, the tsunami in Sri Lanka was a like surging high tide that would not stop. This was because the waters off Sri Lanka are very deep, while those off Thailand are very shallow. The waves that hit Sri Lanka were widely spaced and kept coming for more than an hour. In some cases, they surged far inland, then withdrew far out to sea and then surged again. In many ways, the most dangerous aspect of these waves was the incoming and outgoing current that swept people away and, in many cases, far out to sea. The area around Kalmunai, on the east coast of Sri Lanka about 50 kilometres south of Batticaloa, was the worst-hit area in Sri Lanka. Around 10,000 people were killed along a single 6.4-kilometre section of beach, with entire villages disappearing with hardly a trace left behind. Around 3,000 people were killed around Mullaittavu in northeast Sri Lanka and the fishing villages around it. Entire extended families were wiped out. People were pushed hundreds of meters inland and then pulled out to sea. The beach road and many of the houses that were on it vanished. More than 4,000 died in the southern historical resort town of Galle. Many died where water was funnelled through a busy bus and train station. In Hambantota, 4,500 people died; more than 2,500 bodies were fished from the lagoon near the town. (Hays, 2008)

The Sinhalese areas in the densely populated south, and in the not-so-populated southeast, suffered severely. But the northeast under the control of the Liberation of Tamil Tigers Eelam (LTTE) was even more badly affected. Whether it was the east coast, with its large Muslim population or the southeast with Muslims, Tamils and some Sinhalese population, all suffered loss of lives, livelihoods and income. After some early success in helping people to cope with the devastation, government rehabilitation policies were seen to be inadequate and lacking in focus (De Silva, 2012: 176–177).

The political milieu in Sri Lanka in 2004

The tsunami met the political whirlwind that had already engulfed Sri Lanka and had decisive ramifications on post-disaster strategies. By 2004, the country had already been jostling with ethnic insurgency for over two decades and battling civilian casualties. The government of Sri Lanka, under the United National Party (which came to power in 2001), was facing an unprecedented challenge, wherein the party's leader, Ranil Wickremansing, formed a United National Front (UNF) government with other stakeholders like the Tamil National Alliance and Sri Lanka Muslim Congress. Thus, the prime minister and the president represented opposing parties. UNF's accession to power witnessed the signing of the Ceasefire Agreement (CFA) with the LTTE and the resumption of peace talks. The LTTE even gave unexpected signs of abandoning its long-standing demand for a separate state and appeared to be prepared to accept regional autonomy and regional government. However, the political tides turned once again as the unsteady coalition called United People's Freedom Alliance defeated the UNF in the parliamentary elections held in April 2004 and Mahindra Rajapakse (a senior member of the UPFA and former fisheries minister) was declared the president. The new government was keen to continue with peace talks with the LTTE, but the latter did not reciprocate (De Silva, 2012: 172–175).

However, as the tsunami struck, the government was provoked to collaborate with the LTTE to share the billions of dollars in aid that was given by the international community after the catastrophe. However, incidents on the ground in aid distribution soon reflected the deep distrust between the two establishments. Following this, the government handed over relief distribution to the Sri Lankan army, which the LTTE refused to allow into its arenas. The differences also gave a reason to the LTTE to complain that the government was not providing enough relief to the Tamil areas of the north and the east. (Muni, 2006: xxx) Moreover, according to a report by Human Rights Watch, from 1 February 2002 to 31 December 2006, the Sri Lankan Monitoring Mission (SLMM) reported over 4,000 violations of ceasefire, many of which involved targeted killings and other violence and intimidation against civilians, with the vast majority being committed by the LTTE (Human Rights' Watch, 2007). Rajapakse was constrained, as the coalition-building exercise with the Janatha Vimukthi Peramuna (JVP; the radical Sinhala outfit) was proving exceedingly difficult. It was opposed to the LTTE and, throughout the entire period of its association with the administration, it undermined every government effort to devise a policy of accommodation with the LTTE as part of the existing peace process. Moreover, as the government proposed to launch the Post-Tsunami Operations Management Structure (P-TOMS), which could bring officials and the LTTE together in the reconstruction of areas damaged by the tsunami, there was nationwide controversy that P-TOMS might turn into the basis for a separate state for the LTTE (De Silva, 2012: 177).

The politics of aid and reconstruction

A reconnaissance team comprising American and Sri Lanka experts meticulously published eyewitness reports of the situation immediately after disaster struck. Special arrangements with the Tamil Rehabilitation Organisation (TRO) allowed this team to enter the Tamil-controlled area and carry out a comprehensive assessment of local relief and reconstruction efforts. According to the report, five months after the tsunami disaster, most people from affected communities were still living in temporary or semi-permanent shelters. The heavy winds and long rainshowers of the monsoon season had worsened life conditions for the tsunami victims who had not been able to find permanent homes and were still living in temporary camps. In some instances, temporary shelters were built on sites next to the foundations of destroyed houses. In most cases, however, these shelters were being constructed in resettlement areas. Some of the shelters built were poorly adapted to the hot, humid and wet climate of coastal Sri Lanka (Khazai et al., 2006).

Another report observes that the tsunami and ensuing relief efforts precipitated a renewal of fighting and the death of the Ceasefire Agreement which had been signed between the government and the LTTE in 2002 (following Norwegian mediation). Over two-thirds of tsunami damage was concentrated in Northern and Eastern Provinces, home of much of the Tamil and Muslim minorities. Estimation of the ethnic distribution of damage using community-level data suggests that 48 percent of destroyed homes belonged to Tamils, 23 percent to Muslims and 29 percent to Sinhalese. A similar proportion of death happened among Tamils. The reason for this has been traced back to the demographic reshuffling that both the government and the LTTE were pursuing since the mid-1990s to strengthen their regional bargaining positions. Little did they realise that the complex political ecology and history of population transfer in Eastern Province was pushing large Tamil communities to the coast. In majority-Muslim Ampara, the most heavily affected district, only 30 percent of the population was Tamil, yet 52 percent of destroyed homes belonged to Tamils. Eastern Province Tamils once constituted the principal base of financial, logistical and military support for the LTTE. Ironically, the comprehensive population transfer policies of the LTTE had forced out the Muslims of Northern Province long before the tsunami, rendering a substantial Tamil impact.

The Sri Lankan government received donations worth $2 billion (approximately) from international sources. Throughout Europe, North America and East Asia, vast sums were raised to support the relief and rehabilitation effort. Thus, the British-based DEC raised around 350 million pounds before the appeal for relief aid closed in March 2005. The aid from the United States government focused on reconstruction. The American military was marginally present because the Indian military was proactive. India deployed 14 ships, nearly 1,000 military personnel and several dozen helicopters and planes. Medical teams arrived within hours after the disaster

(Stirrat, 2006). The first ships were in Sri Lanka the day after the disaster, with soldiers digging latrines in relief camps and ships helping to clear sunken ships and debris from the harbours. The World Bank simultaneously restructured ten ongoing operations and made $75 million available for immediate emergency recovery needs, which was quickly approved as the Tsunami Emergency Recovery Program (TERP). The World Bank's support to the government of Sri Lanka directly after the tsunami was to mitigate the immediate suffering resulting from the devastation; to help regain the affected population's livelihoods, restore basic services to them; and to initiate the recovery and reconstruction process (World Bank, December 23, 2014).

Relief efforts of foreign agencies and governments included setting up water purification centres, temporary housing and trauma-counselling centres. However, the relief agencies faced a very different problem at the ground level. It was difficult in eastern Sri Lanka to use large-scale water purification equipment because the tsunami-affected population was spread along the coast rather than focused around one central point. Elsewhere, including the southwest of Sri Lanka, basic amenities such as food and water were never far away from the area of destruction. Thus, they had to concede that the local people and the government would lead the immediate response to the disaster. The 'civil society' came forward with a bold response. Even though road and rail connections were badly hit, individuals and groups from Colombo and inland areas carried food, transportable shelter and medical equipment to the affected.

When the tsunami struck, the president was out of the country, and this compromised the initial involvement of the government in coordinating or managing a bold response to the disaster. But the government paid cremation expenses. Allocation of land for building new homes ran into controversies and legal difficulties. This paved the way for the intervention of foreign organisations, which became involved in the relief effort very quickly. Some of these, such as Oxfam, Save the Children and CARE, were already active in Sri Lanka, while other organisations, not just NGOs but also the military from various nations, arrived within a few days. Gradually, the Sri Lankans became less active; local organisations including NGOs were reduced to the status of junior partners dependent on foreign agencies. Many of the larger NGOs were aware that they lacked the capabilities or capacities to launch activities in Sri Lanka and recognised the need to work with local organisations which had the necessary skill and experience. Yet these were scarce and often already in some sort of relationship with foreign NGO donors. Another problem was that due to the presence of TV teams and reporters, the international and foreign-based NGOs had to spend the money on 'visible' initiatives such as distributing new fishing craft or constructing housing rather than less visible or more indirect forms of disaster relief, for instance, rehabilitation of government offices destroyed by the tsunami. Also, different national branches of

NGOs followed different strategies, and this led to ostensible competition (Stirrat, 2006).

Economically, forecasts of growth dropped from 4.8 percent to 2.8 percent after the tsunami. Since the economy draws heavily from tourism, it was expected that growth would be reduced by 2 percent. Currency and the stock market were not affected much. Fishing was hit hard by the tsunami. Many fishermen had no insurance. However, tsunami aid helped the economy. The central bank was able to lower interest rates. The large amounts of dollars pouring into the country improved Sri Lanka's balance of payments, curbed the rate of inflation and raised the value of the Sri Lankan rupee. Some countries allowed Sri Lanka to delay its loan payments. There was a surge in construction and production of raw materials like cement and steel (Hays, 2008).

Taken together, the tsunami and the relief effort constituted at minimum a net 150 million dollar transfer to the southwest of the country relative to the northeast. The LTTE was bitterly challenged to extract resources and recruit manpower at all costs. Thus, a related impact of the tsunami-reconstruction efforts was to plunge the country into the politics of internal war. This was evidenced in the annual increases in defence spending amounting to 40 per year by 2006. Moreover, there were reports that the government misused the tsunami-relief fund to meet the burdens of new expenditures (Kuhn, 2009). As Prof. Uyangoda observed at that time, the argument surrounding just who is entitled to relief funds from international donors became the subject of such an intense and bitter national debate that a number of international charities stalled the flow of help to the victims of the disaster. Thus, at the local levels, there was a lot of disaffection over the distribution of aid, and many people were clearly disenchanted with the democratic process (McDougall, 2005).

Reflecting on the questionable approach of the government, it has also been observed that the expenditures in the housing sector were considerably higher in Sinhala-majority Southern and Western Provinces than in Northern and Eastern Provinces, particularly in Ampara and Batticaloa District. President Mahinda Rajapakse's anti-ceasefire support base and the virulently nationalistic JVP were perceived to be quite simply indifferent to the larger Tamil cause. Rajapakse's home district of Hambantota received four times the average expenditure per home and six times that received by the most heavily affected eastern district. Donor allocations strategised by the government also betrayed pro-Sinhala bias across bilateral, multilateral and non-governmental donors. Bilateral donors tended to exert a strong bias against areas of LTTE control, while multilateral donors balanced these tendencies. Funding biases resulted in divergent rates of housing reconstruction. By the end of 2006, only 20 percent of homes had been rebuilt in Ampara, 30 percent in Batticaloa. In Southern Province, Hambantota had received more than three new houses for every one destroyed (4,065 built versus 1,290 destroyed), and Matara had almost reached the break-even point, while the opposition stronghold of Galle had seen only 50 percent of homes replaced.

From disaster to diligence; concluding comments

It cannot be overemphasised that the war-torn, ethnically vulnerable and politically volatile island country was devastated by the unprecedented rage of nature. As the previous discussion attempted to highlight, a knee-jerk reaction to the tsunami was the channelling of aid and donations. However, relief and reconstructions are long-term processes and more acutely so in a conflict/insecurity-ridden society. Even more than a decade after the tsunami, Sri Lanka is frequently on high alert for possible earthquakes and tsunamis hitting the coast. At the level of civil society, families of the deceased hold memorials from time to time. Journalists and reporters have documented these in audio-visuals so that neither alertness nor the memories of the devastation die down. Thus, people live in fear, knowing how much damage another tsunami may cause. While at one level, the fear itself acts as a contributor to mental preparedness, it cannot be denied that a proactive role has to be assumed at two key levels: disaster prevention and disaster management.

For disaster prevention, the roles of science and technology, seismology and weather pattern predictions are crucial. Certain mechanisms have already been put in place. The United Nations set up an Indian Ocean Tsunami Warning System. Accordingly, warnings are sent to the National Tsunami Warning Centres that act as a focal point in case of an emergency in countries of the Indian Ocean rim. Also, there are 'no build zones' within 200 metres of the coastline which have been increased to 300 metres recently. Certain *tsunami drills* are conducted from time to time to test preparedness. For example, in September 2018, 28 countries, including Sri Lanka, took part in the annual international tsunami drill. Constant monitoring and evaluation are crucial for averting the onset of a disaster. A more long-term prevention mechanism may be stalling the rapid patterns of climate change and environmental degradation. For this, environmentalists have suggested mangrove reforestation as a fruitful strategy.

At the level of disaster management, the lessons learned from the 2004 tsunami show that local governments must be adequately empowered and equipped for disaster responses. There needs to be effective social and civil society mobilisation to implement management techniques and prevent the escalation of the crisis. In reconstruction efforts, homeowners must be recognised as legitimate stakeholders in the process of allocation of land, use of materials and so on. Grievance redressal cells could operate in different provinces to account for malpractices by either government officials or aid-givers. Another key strategy could be coordination among donors. This in itself could operate at the levels of foreign NGOs existing through local chapters in the country and coordinating among themselves and outsider agencies offering aid and assistance. Above all, to uphold best practices, experts and the government must develop comprehensive manuals for disaster prevention, management and relief strategies.

It must be added that notwithstanding Sri Lanka's specific vulnerabilities to climate changes and disasters as an island country, the countries in South Asia could very well take cues from the experiences gained and lessons learned in the 2004 tsunami. Keeping climate change and disaster management as a top agenda item on a regional level through multilateral forums would go a long way in reiterating the need for awareness, monitoring and preparedness regarding environmental disasters. The Sri Lankan episode is also a pointer to the fact that apart from taking precautionary measures so that anthropogenic activities do not aggravate climate change and natural disasters, South Asian countries must also have prior plans about how to deal with a calamity once it happens and even a politically decided agenda about how to allocate resources received as aid to ensure relief and rehabilitation after tragedy strikes. South Asia, as a part of the Global South, is at once the sufferer and bearer of the costs of industrialisation-induced climate changes. Rising up to the challenge with a well-developed kit of long-term and immediate prescriptions and responses should be accorded 'natural' priority in the policies of these nations.

References

Buzan, Barry, Waever, Ole, and de Wilde, Jaap. *Security: A New Framework of Analysis*. Boulder: Lynne Rienner Publishers, 1998.

Chatterjee, Aneek. *International relations Today: Concepts and Applications*. New Delhi: Pearson, 2010.

De Silva, K. M. *Sri Lanka and the Defeat of the LTTE*. Colombo: Vijitha Yapa Publications, 2012.

Hays, Jeffrey. 'Great Tsunami of December 2004 in Sri Lanka: Damage, Eyewitness Accounts and Rebuilding the Economy.' *Facts and Details*, 2008. Available at: http://factsanddetails.com/asian/cat63/sub411/item2545.html#chapter-0 (Accessed on 21 January 2019).

Heywood, Andrew. *Political Ideologies: An Introduction*. New York: Palgrave Macmillan, 2002a.

Heywood, Andrew. *Politics* (2nd ed.). New York: Palgrave Macmillan, 2002b.

Human Rights' Watch. 2007. Available at: http://hrw.org/report/2007/srilanka0802/2. htm (Accessed on 10 November 2013).

Hyndman, Jennifer. 'The Securitization of Fear in Post-Tsunami Sri Lanka.' *Annals of the Association of American Geographers*, 97(2) (June 2007): 361–372.

International Union for the Conservation of Nature. 'After the Tsunami: Material for Reconstruction – Environmental Issues, Series on Best Practice Guidelines.' Information Paper No. 3, 2005.

Khazai, Bijan, Franco, Guillermo, Carter Ingram, J., del Rio, Cristina Rumbaitis, Dias, Priyan, Dissanayake, Ranjith, Chandratilake, Ravihansa, and Jothyanna, S. 'Post-December 2004 Tsunami Reconstructionin Sri Lanka and Its Potential

Impacts on Future Vulnerability.' *Earthquake Spectra*, 22, No. S3, Earthquake Engineering Research Institute, June 2006, pp. S829–S844.

Kuhn, Randall. 'Tsunami and Conflict in Sri Lanka.' Policy Report, March 20, 2009.

McDougall, Dan. 'After War, Tsunami and Corruption, They Won't Vote for Anyone.' October 30, 2005. Available at: www.theguardian.com/world/2005/oct/30/tsunami2004.Srilanka (Accessed on 21 January 2019).

Muni, S. D. 'An Overview.' In Raman, Moorthy and Chittaranjan (Eds.), *Peace without Process: Sri Lanka's Dilemma*. New Delhi: Observer Research Foundation, 2006.

Stirrat, Jock, 'Competitive Humanitarianism: Relief and the Tsunami in Sri Lanka.' *Anthropology Today*, 22(5) (October 2006): 11–16. Available at: www.jstor.org/stable/4124479 (Accessed on 26 July 2018).

The Guardian. 'Thousands Killed in Asian Tsunami.' December 26, 2004. Available at: www.theguardian.com/environment/2004/dec/26/naturaldisasters.climatechange (Accessed on 21 January 2019).

Tudor Silva, Kalinga, 'Tsunami Third Wave and the Politics of Disaster Management in Sri Lanka.' *Norwegian Journal of Geography*, 63(1) (2009): 61–72. https://doi.org/10.1080/00291950802712145 (Accessed on 21 March 2019).

Vinod, M. J. 'From Traditional to Non-Traditional Security: Building Constituencies for Change in Asia.' In T. Nirmala Devi and Adluri Subramanyam Raju (Eds.), *Envisioning a New South Asia* (pp. 143–154). New Delhi: Shipra Publications, 2009.

Waever, Ole. 'Securitization and Desecuritization.' In Ronnie D. Lipschutz (Ed.), *On Security* (pp. 1–31). New York: Columbia University Press, 1995.

Waldman, Amy. 'Thousands Die as Quake-Spawned Waves Crash onto Coastlines across Southern Asia.' *New York Times*, December 27, 2004. Available at: www.nytimes.com/2004/12/27/world/asia/thousands-die-as-quakespawned-waves-crash-onto-coastlines-across.html (Accessed on 21 January 2019).

World Bank. 'Sri Lanka at a Glance.' Report, 2005. Available at: www.worldbank.org/cgibin/sendoff.cgi?page=%2Fdata%2Fcountrydata%2Faag%2Flka_aag. Pdf (Accessed on 13 July).

World Bank. 'Lessons Learned from the Sri Lanka's Tsunami Reconstruction.' Report, December 23, 2014. Available at: www.worldbank.org/en/news/feature/2014/12/23/lessons-learned-sri-lanka-tsunami-reconstruction (Accessed on 21 January 2019).

Part IV

Adaptation strategies to combat the challenge

From theory to praxis

10 Ethical dilemma of sustainable development as combat mechanism to climate change

Sarbani Guha Ghosal

According to reinsurer Munich Re, 2017 was the costliest year to date in terms of natural disasters, with total damage estimated as US$310 million (Walsen et al., 2018). The issue of natural disasters is closely linked to the issue of environmental challenge, which has been with us since our planet came into existence about 4500 million years ago. The emergence of human beings and their civilization has changed the history of the planet because since inception we have not remained a product of the environment but its active transformer. The progress of human imprints has coincided with environmental decay since pre-historic times. During the past 2 or 3 million years, the major agents of environmental change have been climate and humans, and these two are intertwined in a complex way, directly affecting the processes operative in the environmental system in multiple ways.

In the twenty-first century, at the policy-making level, we have come to acknowledge the evils of human-generated or anthropogenic environmental problems. Before the last two decades of the earlier century the dominant ideology of the west was unrestrained development with fullest possible exploitation of natural and human resources. The notion of 'human exemptionalism' (Dunlap and Catton, 1994) overcoming all natural constraints was the credo of that period. Very precisely, it is in the last quarter of the earlier century that the dangers of environmental decay and climatic challenge began to perturb us both at the individual and collective level. The concept of human exceptionalism started generating resentment; natural scientists, policy-makers, opinion leaders and the general public gradually became vocal about pollution and environmental problems. Society's dependence upon natural resources and the need to protect the environment came into focus. Different disciplines like environmental sociology, environmental politics, environmental economy and environmental psychology emerged with the intention to facilitate environment-conscious policy-making.

We have noted that global warming, among all the environmental issues, threatens the well being of developed and developing countries most. Naturally, mitigation of climatic challenges needs to address the issue of global warming with the greatest priority. Like many other issues of international politics, global warming, climatic change and environmental decay have also developed a clear north-south divide. The advanced west was prompt enough to shift the entire onus of environmental decay and climatic destabilization onto poor and developing nations, particularly on India, China and some African and other Asian countries. This north-south divide is consequently manifested with respect to environmental policy-making with lots of statistical and scientific armour.

The idea of sustainable development, instead of a dominant scientific and industry-based development paradigm rooted in nonstop conquering of nature associated with unending expansion, production, consumption and unrestrained growth, started to become popular in the 1980s. The traditional notion of growth gave way to sustainable development, and growth became green growth (Shiva, 1991). After many deliberations over a considerable period of time, ultimately, world leadership agreed to accept environmental issues as 'common but differentiated responsibilities' at the Earth Summit of Rio de Janeiro, 1992. The Common But Differentiated Responsibilities (CBDR) principle was formalized with its inclusion in United Nations Framework Convention on Climate Change (UNFCCC), article 3, paragraph 1, after continuous insistence from the developing bloc. The idea of sustainable development has emerged as a major political and environmental discourse as a combat mechanism to deal with environmental decay and climatic challenges. However, it has been recommended in particular to developing countries as a development path that would not replicate the environmental degradation that has been incurred in industrialized countries. Quite naturally, leaders from the developing world had a different agenda, particularly during the turbulent world political scenario of the 1980s and 1990s, and they were not ready to accept the western concept of sustainable development or to bear the entire load of keeping the world safe for posterity. The developed western world has accumulated wealth and prosperity by making use of nature as well as by mercilessly exploiting the natural resources and environment of the developing states. So, it was now seen as hypocritical of the former to ask the latter to protect the environment and control populations at the expense of their chance at development (Grainger and Purvis, 2005).

Explaining the concept of sustainable development

From a theoretical point of view, the concept of sustainable development is generally considered a comparatively new entrant in the domain of social studies. A humane ecological perspective actually lies at the core of it. It emphasizes breaking the pre-eminence of human exceptionalism, drawing clues from the work of ecologists, environmentalists, natural scientists and philosophers to highlight the different kinds of environmental problems and to identify the basic material conditions of societies, like population, size of the state, growth level, scale of production and consumption and use of technology as the primary factors responsible for those problems. We have to remember, despite the fact that sustainable development is a comparatively new concept, that ideas of intergenerational equity and respect for the interest of coming generations are not new. Hinduism, Islam, Taoism, Christianity, Confucianism and Buddhism, like all major religions, preach the need for restraint in consumption as an essential human virtue and a way to reach god. It is worth mentioning here that as early as 1798, Thomas Malthus, one of the pioneering theorists of capitalist economy, pointed out the dangers of exhausting environmental resources. He referred to the fact that unchecked population growth would inevitably exhaust earth's available resources. His social position led him to believe in preventive checks like late marriage, homosexuality and forms of birth control, and for the poor population, natural checks like starvation, disease and war would be controlling factors (Flint, 2011). In the Marxian framework of political economy, the idea of a 'metabolic rift' gave adequate warning for impending environmental crisis. Marx saw an emerging ecological crisis developing out of the degradation of soil resulting from the exploitative practices of capitalist agriculture, which robbed the soil, as it did the workers, for profit maximization. Marx argued that rapid urbanization driven by capitalist industrial development separated people from the land and thereby created a rift in the metabolic exchange between people and nature that had sustained soil in pre-industrial times. In pre-urbanized societies, nutrients taken from the land in agricultural production were recycled back into the soil; contrarily, in urbanized societies, nutrients were shipped to the city in the form of food and fibre, where they became a waste problem. Waste in the city, instead of enriching the soil, went into the dumps and sewers and subsequently polluted rivers and streams. Thus, Marx saw the capitalist urban-industrial society as fundamentally unsustainable, which is a main reason for his advocacy for even distribution of people across the landscape. Marxist 'metabolic rift' theory highlights how societal-environmental relationships can be unsustainable even before they push up against the limits of scale. Later on, this theory inspired others to generalize it as a conceptual apparatus to examine the global carbon cycle and climatic change and the oceanic crisis stemming from the degradation of fisheries or nitrogen cycle issues. It focuses on the fact that capitalism has inherently unsustainable features and hence calls for reorienting the basic objectives of production. The Marxist political economy actually induces the idea of a 'climate resilient' economic policy, reducing population growth and economic activities together following the prescription of ecologists. They suggest that in order to overcome the modern environmental crisis, societies need to halt the treadmill production process and the demand for endless growth and private profits and move to an economy that caters to human needs (York and Dunlap, 2016). This

Marxist idea synchronizes very well with the words of India's celebrated Mahatma Gandhi,

> there is enough for everyone's need, but not enough for everyone's greed.

This clearly means that growth is never an end in itself. Value-laden practices towards growth are of the utmost importance. Growth is necessary but cannot be the sufficient condition for human development unless adequate measures are taken for removing poverty, ill health, educational and housing problems and other forms of human drudgeries. Efficient delivery of public goods and services and infrastructural development backed by suitable government procedures and regulations are of utmost need.

Sustainable development vis-à-vis climate change

Now let us discuss in brief the concept of sustainable development as we understand it today, as a combat mechanism to environmental degradation and climate change, and gradually we will move towards the dilemma ingrained in the idea. The first caution to be remembered is that the concept neither has any universal acceptable version, nor has it any definite boundary. Almost 80 different, often competing and sometimes contradictory definitions have been identified. The concept can easily be termed an 'environmental paradox', for nearly all commentators disagree on what is demanded from the earth and what the earth is actually capable of supplying. Thus, in the ultimate formulation, sustainable development is a 'notoriously difficult, slippery and elusive concept to pin down'(Williams and Millington, 2004). However, the core task before global sustainable development is to respond positively to climate change for today and the future, so for this action-oriented concept instead of a 'good definition' a 'good description' is required (McNeill, 2000). Moreover, the scope of sustainable development has increased over time. Starting with a mere technique of preservation of almost-extinct or depleted biological species, it was later expanded to the conservation and preservation of all of biodiversity as well as formulating an alternative form of overall development, not environmental alone. At present, thus, when we talk about sustainable development, we specifically refer to three interconnected areas of concern – environment, society and economy. Simultaneously, we have to keep in mind that sustainability is never an end state, but it is reinvented continuously. In the introductory note of their comprehensive collection on sustainable development, Agyeman et al. (2003) observed that our emphasis on sustainability mostly emphasizes precautions, the need to ensure a better quality of life for all, now and into the future in a just and equitable manner while living within the limits of supporting ecosystems. Very categorically, they have stated that unless analyses of development begin not with symptoms and environmental and economic instability but with the cause, social injustice, then no development can be sustainable. We have to remember that

> a truly sustainable society is one where wider questions of social needs and welfare and economic opportunity are integrally connected to environmental concerns.
>
> (Agyeman et al., 2003)

In other words, it is not wrong for us to conclude that sustainable development as a combat mechanism against climatic challenges genuinely addresses all possible determinants of the problem. The publication of the report 'Our Common Future' (2018) from the World Commission on Environment and Development (WCED) of UNO, popularly known as the Brundtland Report (1987) changed to a great extent the international attitude and orientation towards the issue of climatic change and its corrective measures. Instead of focusing on the traditional conservation-based notion of sustainability as developed by the International Union of Conservation & Nature (IUCN) in the 1980s, this report adopted a more comprehensive and holistic notion of sustainable development. It developed a framework with emphasis on the social, economic and political context of development. The term 'development' simply refers to 'what we all do in attempting to improve our lives'. Side by side, the report categorically states,

> we must ensure development meets the needs of the present without compromising the ability of future generations to meet their needs . . . sustainable development is not a fixed state of harmony, but rather a process of change in which the exploitation of resources, the direction of investments, the orientation of technological development and institutionalized change are made consistent with both future and present needs.

The report makes it clear that sustainable development is neither any easy method nor a straightforward process; often, painful choices have to be made, and in the final analysis, it 'must rest on the political will'. The World Summit of Sustainable Development (WSSD), 2002, held at Johannesburg, pointed out five interconnected priorities to climate change: water, energy, health, agriculture and biodiversity. The discussion that we have provided clearly shows that sustainable development is never merely a normative concept, and its practical side is of utmost importance and can be identified as a major praxis of our century. It is precisely from this point that several critiques of the concept have emanated. The main theme of the idea is,

> a fair distribution of natural resources available among different generations, as well as among the populations of the First, the Second and the

> Third world of our own generation. Though the concept is given massive support throughout the world its realization is highly problematic. The greatest problem is the operationalization of the concept and from it arise a number of unanswered questions that need to be addressed.
>
> (Dietz and Straaten, 1993)

It clearly appears that from the theoretical point of view, the sustainability concept has to be mainstreamed as an integral part of climate challenge–related policies. Here there is a need for clear understanding and practices of several mitigation technologies and their relationship with the state's overall development direction. Non-climatic policies cannot be excluded, and, most importantly, regular stocktaking of renewable and non-renewable resources should be done. However, attainment of none of these is debate free. The entire mechanism of sustainable development with the intention to protect the environment from depletion and degradation is marked by several interrelated dilemmas. In this chapter, the focus will now be on some of the major interpretations like the weak and the strong concept of sustainable development or the environmentalist and ecologist perspective of sustainable development; the northern interpretation versus southern interpretation; and the green, the nationalist and the feminist interpretations, all of which in reality make sustainable development an unputdownable combat mechanism towards climate change as a whole.

The weaker and stronger concepts of sustainable development actually developed while resolving the conflict between the demand on the environment and the resources of the environment. Weaker sustainable development takes a completely anthropocentric view and finds nothing wrong in following the prevalent pattern of economic development (Williams and Millington, 2004). It believes in separation of people from nature and that people have the right to dominate nature by using it as a mere resource for the benefit of the society and individual. It considers that capitalism can very well accommodate environmental problems and climate challenges. It has enough faith in the technological progress of capitalism to ultimately manipulate and solve all kinds of environmental problems. A section of these theorists believe in the idea of ecological modernization, which refers to the possibility of improving the efficiency of economic growth by using fewer natural resources and by adopting sustainable waste practices. There is a second group which emphasizes the idea of environmental justice (Agyeman et al., 2003).

They speak of the importance of economic growth and also highlight the need to redistribute the cost and benefits of development in a more equitable manner on either the intra- and/or intergenerational level. They believe certain resources can be depleted as long as they can be substituted by others over time. They believe that natural capital can be used up as long as it is converted into manufactured capital of equal value. However, the problem of this is that often it is difficult to assign a monetary value to natural materials and services, and some of these can never be replaced by manufactured goods and services.

The stronger sustainability discourse completely tries to abolish anthropocentrism and specifically states that renewable resources also must not be used faster than the rate of their replenishing. It emphasizes changing the nature of demands made on the earth. They criticize their opposites as a weaker version of sustainable development more about sustaining development than sustaining the environment, nature, ecosystems or earth's life support systems (Williams and Millington, 2004). This group of scholars advocates for a small-scale decentralized way of life based upon greater self reliance so as to create a social and economic system that is less destructive towards nature.

The environmentalism-ecologism dilemma is a major concern in the domain of environmental crisis and subsequent climate change. Environmentalism is often considered a superficial and reformist vision that regards environmental problems as management problems and believes resolution can be found in the context of the existing dominant political and economic framework 'without any rigorous change in our values and culture' (Achterberg, 1993). Contrary to this, ecologism or deep ecology proposes basic structural change. It calls for a radical change in our attitude towards nature and natural problems and thus in our political and social systems as well. Environmentalism is an anthropocentric ideology, whereas ecologism is fully ecocentric. The environmentalist and ecologist both are quite concerned about environmental decay, but their proposed remedial strategies vary wildly. We have to remember here that in reality, ecocentrism does not mean an entire subordination of human values to those of nature but equal recognition of non-human intrinsic values. It is from this point that the issue of sustainability or sustainable development derives as the correct combat mechanism to environmental decay and climate change.

The green view explains the concept of sustainable development in a somewhat different light. Drawing clues from the ecologist point of view Andrew Dobson, eminent British political scientist and green theorist, holds the view that environmentalism follows a managerial approach towards environmental challenges with a belief in mitigating and solving them within the existing socio-political and economic parameters – in other words, without fundamental changes in present values and patterns of production and consumption. On the other hand, ecologists believe that a sustainable and fulfilling existence presupposes radical changes in our relationship with the non-human natural world and in our mode of socio-political life. Green interpretation pledges for a 'non-violent revolution to overthrow our materialistic industrial society and in its place tries to create a new economic and social order which will allow human beings to live in harmony with the planet'. Dobson says that the character of the green movement, which is actually the most radical and important political force since the birth of socialism, will be completely missed if we restrict our understanding to the guise of an environmentalism that seeks a 'cleaner service economy sustained by cleaner technology and producing cleaner affluence' (2003).

Ecologists and greens are of the opinion that a sustainable society should have two major aspects. First, in the advanced industrial countries, consumption of material goods has to be reduced, because a limit to growth is impossible without limit to consumption. Side by side, it is also to be remembered that mere economic growth cannot satisfy all human needs. Deep greens believe that a sustainable society is always a spiritually fulfilling society. Dobson is of the opinion that a

> radical green programme can hardly be understood without reference to the spiritual dimension on which it likes to dwell.
>
> (ibid.)

It is precisely in this point that policy-makers dealing with the issue of environmental and climatic change launch staunch criticism against the deep greens, because since the Renaissance and Enlightenment, the domain of spiritualism has been separated from political life, and spiritualism can never be a mandatory principle of human existence.

Deep ecologism and the greens are equally critical of both the capitalist and socialist positions towards climatic change, because both ideologies glorify utmost industrialization. Actually, the difference in the ownership of the means of production has not made any spectacular difference in their attitude towards environmental degradation. Moreover, the high level of environmental damage caused by the erstwhile East European nations, particularly after the effects of the Chernobyl disaster of 1986 became public, was no less harmful than those caused in capitalist economies.

Besides the issue of society and economy, the idea of sustainable development has given birth to a considerable number of dilemmas in the domain of political theory and political systems as well. Actually, the landscape of political theory and practices has transformed enormously owing to the ecological challenges of the last few decades. Virtually no branch of political theory has escaped the influence of sustainability. Andrew Dobson, in this context, has raised some very important questions relating to the issue of compatibility between environmentalism and ecologism vis-à-vis the existing model of liberal democracy and its institutions. Dobson is of the opinion that environmentalism and liberalism are compatible, but ecologism and liberalism are not, because much of liberal political theory runs counter to radical environmental ideology. Individualism, the pursuit of private gain, limited government and freedom of the market are clearly contradicted by radical ecological commitments (ibid.). A more moderate theorist, Achterberg, predominantly belonging to the domain of environmentalism, comes forward in resolving this dilemma when he says that the political conception of sustainable development is largely based upon central elements in the political philosophy of liberalism. Therefore, it can play a legitimizing role within a liberal democracy, which is necessary in view of the radical changes connected with the solution to or control of the environmental crisis (Achterberg, 1993). The philosophy of socialism by criticizing capitalism as the source of all the ills of contemporary society has failed to fulfil the compatibility criterion with ecologism, as the latter condemns industrialization per se as the contaminating force. According to Dobson, the green are actually 'beyond left and right'. In his later writings, Dobson dealt with another related issue (2003), the relationship between social justice and environmental sustainability. No doubt this issue is filled with enormous dilemmas. Initially, Dobson was of the opinion that these two are not always compatible objectives, though from a 'political point of view there are tremendous benefits in marrying the two'. On observing the German situation, he concluded that any kind of rapprochement between these two can be only temporary and transient. To him, the reds and greens, meaning the advocates of social justice and ecologism, respectively, have intrinsically different objectives, and to expect socialists and environmentalists to form a common cause is as unrealistic as to expect liberals and socialists to make a common cause. Their difference is not merely tactical but strategic as well. At the same time, he believed that social justice is as contested a term as sustainable development. However, later on, mellowing to an extent, he stated that from the point of view of providing future generations with adequate opportunity and not depriving them of enjoying and experiencing available alternatives, 'the objectives of social justice, i.e., equal distribution of opportunities and sustainability as the preservation of biodiversity are the same'.

The nationalist approach to sustainable development is another kind of political concept which argues for non-infliction of harm in our environment so that our descendants are not affected in any way. Nationalism considers the idea of the nation as an ongoing chain across generations, so it considers environmental issues a matter of distribution of access to environmental goods across generations and rules out any policy that could arbitrarily inflict harm on future generations Shalit (2006). However, in the domain of practical politics, many strategies adopted are often not congenial to sustainability discourse and often encourage parochialism, hatred towards other communities and destruction of the environment. In this context, we can say that environmentalism and nationalism can find some of their goals and values compatible with civic nationalism (Hamilton, 2002). Particularly regarding refugee rehabilitation, sharing of water and other important resources essential for human sustenance, a sustainable nationalist format may not work and instead can bring serious conflicts between groups and communities by creating artificial boundaries. International strategies, which are often used for mitigating climate change and which are important armour ingrained in the concept of sustainability, often face stringent opposition from the protagonists of economic nationalism and thereby contribute to environmental degradation.

Since the 1970s, the relationship between feminism and feminist politics vis-à-vis environmental challenges and sustainability techniques has become an important area in academics and at the policy-making level as well. Women's relation to the environment, particularly in the countries of the global south, has become an important theme in any global discussion on environmental crises. At the beginning, the patriarchal development discourse was not hesitant to put the responsibility for environmental degradation, particularly depletion of fossil fuels, on poor women of developing countries. This trend was prevalent in the environmental debates of the west as well as the insensitive political leadership of global south, which put the onus of environmental decay on poor women of their own states.

The situation, however, gradually started to change in the following decades. At the Nairobi forum of 1985, held parallel to the UN and Development Conference, women's actions and special role in environment management were presented, with case studies highlighting their involvement in forestry, agriculture and energy, based mostly on experiences from the global south. Women were portrayed as environmental managers whose involvement is crucial in achieving sustainable development. With the publication of the Report of the World Commission on Environment and Development (1987) the Women, Environment, Development (WED) debate focused on transforming the images of poor women of the developing south from mere victims to images of strength equipped with indigenous resourcefulness. A cultural stream of thought has developed emphasizing women's position as closer to nature. It perceives the women-nature relation as one of reciprocity, symbiosis, harmony, mutuality and interrelatedness due to women's close dependence on nature for subsistence needs (Braidotti et al., 1997)

From the different forums of the UN, emphasis is also placed on the theme that women must remain at the heart of climate action if the world is to limit the deadly impact of disasters such as floods and storms. In a very recent declaration Mary Robinson, UN rights commissioner and a former member of the UN climate envoy, stated that women are most adversely affected by natural disasters and yet are rarely put front and centre of efforts to protect the most vulnerable. She has also said that climate change is a man-made problem and must have a feminist solution, because feminism does not mean excluding men; it is about being more inclusive of women and, in this case, acknowledging the role they can play in tackling climate change (Tabary, 2018). As the issue of climate change is more and more associated with the issues of poverty, inequality, migration, local communities, peace and security or human development as a whole, the role of women is becoming more significant, though sometimes in contradictory ways.

It is often said that women can play a noticeable part in combating climatic challenges; particularly in the domain of mitigation and adaptation strategies, they can become important agents. At the same time, the problem with this approach is that it often overlooks the relational aspect of gender; that is, what are the exclusive tasks of men towards climate change, or is there any such real gender difference from the perspective of power structures? Along with this, it almost glorifies vulnerability and caretaking as intrinsic to women, raising them to the status of virtues. It also focuses on individuals instead of social relations and social structures and puts an additional burden on poor women, with more tasks and responsibilities beyond their conventional ones. Furthermore, this kind of neoliberal construction of poor women of the south, predominantly as rational agents of economy, refers to certain kind of political conditioning for enjoyment of their rights and ultimately develops boundaries for social exclusion (Jerneck, 2018).

The north-south controversy of sustainable development has posed not only a serious ethical dilemma but a stern policy-framing dilemma as well. It is often considered an agenda of the developed north to annihilate all kinds of possible challenges and competition from the south. The rise of sustainable development as a popular concept coincides with the rise of the east, particularly after the emergence of the People's Republic of China as a frontal power, which has never been positively accepted in the north. Now, developed countries always put the environment first, ignoring the need for development of the developing world; thus, United Nations Conference on Environment and Development (UNCED) documents refer to it as the optimum path for development. Grainger and Purvis thus believe that it was simply

> a device to reconcile the aspiration of the developing countries to develop and the developed world's desire to curb this development in order to protect the global environment.
>
> (Grainger and Purvis, 2005)

They state that sustainable development has its place only in the minds of states as a codeword for bargaining. The developed world wants a better global environment, while the developing world is for more development, and groups of states will trade off against each other in the course of extended negotiations. This means 'Sustainable development at the global level remains a compromise, but a different and less attractive one from that which most idealists would like'.

The future of sustainable development in particular and the issue of climate change in general become more uncertain when the head of the most powerful state of the north sincerely believes that global warming is just a hoax invented by the Chinese to attack US manufacturing. It is thus often stated that through the mechanism of sustainable development, first world countries that have caused serious environmental degradation for centuries encourage developing countries to reduce pollution. However, pollution generated to date, particularly in areas like the carbon emissions and nuclear waste disposal of western countries, is much higher than the sum total of the developing states. Critics thus argue that for a planet where 20% consume 80% of natural resources, a scheme of sustainable development is just not an acceptable proposition (Brockington et al., 2008).

The north-south debate on sustainable development ultimately poses a serious dilemma by culminating in a nature versus people debate. The north prioritizes conservation of nature over alleviation of poverty, while southerners want to invert the order. Vandana Shiva, the physicist turned environmentalist, is of opinion that G7 can demand a forest convention that imposes international obligations on the so-called Third World to plant trees. But the Third World can neither reverse it, nor can it demand simultaneously that industrialized countries should reduce the use of fossil fuels and energy. Shiva correctly assessed that all 'demands are externally dictated – one way – from North to South. The "global" has been so structured that the North (as the globalized local) has all rights and no responsibility and the South has no rights, but all responsibility' (1991).

It is easy to conclude from this that sustainable development actually establishes dualism and dichotomies. It is highly partial in approaches by suggesting different sets of rules and practices for different groups in different areas. The conflict between life-producing and preserving activities is not resolved either. Anthropological critique also poses certain dilemmas over the concept of sustainable development. Theorists from New Zealand, Rixecker and Matua (2003), have stated that the contemporary system of environmental politics is enshrined in and legitimized through international and national laws premised upon the western legal system and modernity. It becomes especially challenging for indigenous people to reclaim and control their native homes, cultures, practices and beliefs. The rise of the corporate giants has added another dimension to the issue with the emergence of the multinational corporations as the major economic and political power brokers; the place actually was reserved for heads of states and national governments in the past. The indigenous in this context have to be more resourceful than ever before to secure and protect their cultural and environmental heritage (Rixecker and Matua, 2003)

Some other anthropologists view that the emphasis on environmental concerns as held in the sustainable development discourse is inherently technocratic and thereby gives an advantage to etic over emic perspectives in application.[1] The northern version of sustainable development actually emphasizes increasingly bureaucratic environmental institutions where advantages are reserved for those who can make better legalistic negotiations and compromises to manage environmental issues and prefers the creation of top-down projects. It is also said that emic perspectives are often sidelined in sustainable development discourse, and environmental concerns are addressed in such a way that is ahead of the interests and desires of the locals. The entry of foreign aid with donors' preconditions particularly has increased the importance of local elites and power brokers in the application of sustainable development, quite often undermining the local interests and thereby creating more environmental hazards (Smyth, 2011).

India and sustainable development

This section of the chapter sheds some light to the specific case of India and its sustainable development agenda, as this country, along with other South Asian states, is severely facing the brunt of climate change. In the context of fast-changing climate, the corollary climate politics of conflict, inequality, poverty, hunger, rapid urbanization, population growth, ill health and illiteracy are challenging the efforts of developing countries like India in implementing sustainable development goals. The UN Sustainable Development Report of 2018 clearly stated this fact. For the first time in more than a decade, there are now approximately38 million more hungry people in the world, rising to 815 million in 2016 from 777 million in 2015 (Global Hunger Index, 2019). According to The State of Food Security and Nutrition in the World, 2018 Report, almost 195.9 million people are undernourished in India, that is, 14.8% of the population. Also, 51.4% women of reproductive age between 15 to 49 years are anaemic. The Global Hunger Index of 2018 ranked India in 103 out of 119 qualifying countries. India's score here is 31.1, which is identified as a very serious situation in the Report. However, India is world's second-largest food producer, and there are 84 billionaires in India. The FAO report also identified India as the home of the second-largest undernourished population in the world. Frightfully, 38.7% of Indian children under five are stunted, reflecting chronic undernutrition and low height for their age; 15.1% of them are wasted, reflecting acute undernutrition and low weight; and finally, 4.8% die before reaching the age of five. UNICEF in 2016 declared India a land with one-third of the world's stunted children (Global Hunger Index, Case Studies, 2016).

Unfortunately, India is capable of providing food security in the true sense of the term with strong political will and effective distribution policies. But it still faces a long road ahead to achieve zero hunger. In this land of plenty, the top 1% of the population owns more than 50% of the resources. Thus, the national nutritional paradox can only be resolved by challenging the existing social, economic and political structures that deliberately increase vulnerability of the majority of the population by encouraging discrimination of different kinds.

In the context of the previously mentioned situation, we have to remember that the success of sustainable development depends a lot on the bottom-up approach. Diverse and contextual specificities of different cases require special attention, which in Marxian political economy has been identified as 'multiple determinations'. Any kind of sustainability study and practice thus is never confined to livelihood design, energy situation or issue of carbon emission alone. Patterns of social differentiation and social distribution, class formation, gender hierarchies and racial and ethnic discrimination should be explored and addressed properly. Actually, sustainability and development paradigms are 'highly variable, contingent and conjectural'. This has important implications for monitoring and evaluating change. Singular indicators or metrics of transformation will be always insufficient, and a process-based, reflective learning approach is actually essential. This must encompass active reflection on not only changing structural drivers but also shifts in micro-level relations and practices, as well as a deep appreciation of politics (Scoones, 2016).

Constructing pathways to sustainability and development is inevitably a normative struggle rooted in political and moral choices. Focus must be placed on people's agencies and social relationships as well. Different kinds of formal and informal networks, alliances and coalitions connect diverse actors, including the state and business actors – national and multinational, scientific and technical elites, donors and citizen's movements all should be put on the same platform, and all of them have to work amicably to realize sustainable development goals. Similarly, how local knowledge and practices can be connected with wider transformation in national and global strategies must be examined carefully.

India in international relations has now emerged as an important factor in climate diplomacy which is strategically guided by wider foreign policy objectives, on the one hand, particularly in its strategically important relationships with the United States and China and, on the other hand, as a committed partner in the South and South-East Asian region to carry on the battle against climate change in the terrain. The first kind of relationship underlines more of a mitigation approach, while the second one emphasizes adaptation techniques more. However, foreign policy imperatives sometimes contradict domestic policies, particularly related to the underprivileged sections of society. So, how foreign policy objectives and domestic needs can be compromised successfully is always a major challenge for India's sustainable development outlook. Energy security, carbon emissions and economic development are often considered the major issues in the domain of environmental and climate diplomacy for India, but unless these are combined with the issues of food, health and other social discrimination practices, fruitful results in the domain of sustainable development are hard to find. Moreover, the situation becomes more dismal as reports have indicated that implementation of sustainable development goals in India by 2030 will cost around US$14.4 billion (Dubash and Khosla, 2015). The recent trend of cutting financial allotment in social sectors during the last few years has thus created a major dilemma in coalescing the different determinants of sustainable development.

In reality, unless issues of economic growth, ecological deficit and human development predicaments are addressed in a holistic manner, taking into account the different kinds of specificities as well, progress cannot be made in any of the mentioned concerns. Finally, as Amartya Sen emphasized, democracy and human rights are the keys to sustainable development, and good development goals may not always be quantifiable. Almost in the same vein, Joseph Stiglitz said that sustainable development goals are often more important than the millennium goals of UN, as those are required not for developing countries alone but for developed countries as well. Furthermore, sustainable development goals are made not for the governments only but for society as a whole, and an increase in gross domestic product (GDP) cannot be an indicator of reaching sustainable development goals (Victorero, blog, 2015).

Conclusions

It is not wrong to conclude that India must carefully select a multiple development objective framework with concrete implementation actions for the immediate present and coming years. Instead of numerical targets, holistic good should be the aim, and only then can credibility be earned both at the domestic and international levels. In an interactive relationship between domestic policy and international positions, which has significant shifts at different points of time, India is increasingly becoming a testing ground for policies that internalize climate change considerations in development (Dubash et al., 2018).

Last but not least, addressing the global climate challenge and implementation of substantial development goals in the context of substantial development challenges cannot be attained only by political and technological leadership. The multiple challenges involved in accelerating sustainable development and mitigating climate change definitely need positive and sensitive intellectual leadership.

Note

1 *Etic* and *emic* perspectives are two anthropological approaches for understanding social behaviour and culture. The *etic* perspective refers to the perspective of the observer, while *emic* is the perspective of the subject.

References

Achterberg, W. 'Can Liberal Democracy Survive Environmental Crisis.' In A. Dobson and P. Lucardie (Eds.), *The Politics of Nature*. London, UK: Routledge, 1993.

Agyeman, J., Bullard, D., and Evans, B. (Eds.). *Just Sustainabilities – Development in an Unequal World*. London: Earthscan, 2003.

Braidotti, R. Charkiewicz, E., Hausler, S., and Wieringa, S. 'Women, Environment and Sustainable Development.' In N. Visvanathan, L. Duggan, and N. Wiegersma (Eds.), *The Women, Gender and Development Reader*. London: Zed Books, 1997.

Brockington, D., Duffy, R., and Igoe, J. (Eds.). *Nature Unbound*. London: Earthscan, 2008.

Dietz, E. J., and Straaten, J. 'Economic Theories and the Necessary Integration of Ecological Insights.' In A. Dobson and P. Lucardie (Eds.), *The Politics of Nature*. London: Routledge, 1993.

Dobson, A. 'Social Justice and Environmental Sustainability – Never the Twain Shall Meet.' In J. Ageyaman, R. D. Bullard, and B. Evans (Eds.), *Just Sustainabilities* . . . London: The MIT Press, 2003.

Dubash, N.K., and Khosla, R. 'India's Sustainable Development Led Approach to Climate Mitigation for Paris.' Centre for Policy Research, August 2015. Available at: www.cprindia.org/research/reports/india's-sustainable-development-led-approach-climate-mitigation-paris (Accessed on 23 December 2018).

Dubash, N. K., Khosla, R., Kelkar, U., and Lele, S., 'India and Climate Change: Evolving Ideas and Increasing Policy Engagement.' *Annual Review of Environment and Resources* 43(1) (October 2018). Available at: www.annualreviews.org/doi/abs/10.1146/annurev-environ-102017-025809.

Dunlap, R. E., and Catton, W.R. 'Struggling with Human Exemptionalism: The Rise, Decline and Revitalization of Environmental Sociology.' *The American Sociologist,* 25(1) (March 1994). Available at: www.link.springer.com/article/10.1007 (Accessed on 2 December 2018).

Flint, Colin. *Introduction to Geopolitics*. New York: Routledge, 2011.

Grainger, A., and Purvis, M. (Eds.). *Exploring Sustainable Development.* London: Earthscan, 2005

Hamilton, P. 'The Greening of Nationalism: Nationalizing Nature in Europe.' *Environmental Politics,* 11(2) (Summer 2002): 27–48. Available at: www.vedegylet.hu/okopolitika/Hamilton/Greening/Nationalism.pdf (Accessed on 20 January 2019).

Jerneck, A. 'What about Gender in Climate Change? Twelve Feminist Lessons from Development.' *Sustainability* (February 28, 2018). Available at: http://www.mdpi.com/journal/sustainability (Accessed on 20 December 2018).

McNeill, D. 'The Concept of Sustainable Development.' In L. Keekok, A. Holland, and D. McNeil (Eds.), *Global Sustainable Development in the 21st Century*. Edinburgh: Edinburgh University Press, 2000.

Report of the World Commission on Environment and Development, 1987. Available at: www.un-documents.net/our-common-future.pdf (Accessed on 11 November 2018).

Report on Global Hunger Index. Available at: www.globalhungerindex.org/results (Accessed on 2 January 2019).

Report on Global Hunger Index, Case Studies. Available at: www.globalhungerindex.org/case_studies/2016_india.html (Accessed on 2 January 2019).

Rixecker, S. S., and Matua, B. T. 'Maori Kaupapa and Inseparability of Social and Environmental Justice.' In J. Ageyaman, R. D. Bullard, and B. Evans (Eds.), *Just Sustainabilities – Development in an Unequal World*. London: Earthscan, 2003.

Scoones, I. 'The Politics of Sustainability and Development.' *Annual Review of Environment and Resources*, 41(1) (November 2016). Available at: http://researchgate.net/publication/305801984 (Accessed on 2 January 2019).

Shalit, A.D. 'Nationalism.' In A. Dobson and R. Eckersley (Eds.), *Political Theory and Ecological Challenge*. Cambridge, UK: Cambridge University Press, 2006.

Shiva, Vandana. *Ecology and Politics of Survival.* New Delhi: Sage Publications, 1991.

Smyth, L. 'Anthropological Critiques of Sustainable Development.' In *Cross Sections*. 2011 (VII). Available at: www.C:/users/ADMIN/downloads/additional%20readings%20%20anthropology%20and%20sustainability%20(3)pdf (Accessed on 12 December 2018).

Tabary, Zoe. 'Climate Change Is a Man Made Problem with a Feminist Solution.' *Global Citizen*, June 19, 2018. Available at: www.globalcitizen.org/en/content/climate_change_problem_feminist_solution (Accessed on 2 January 2018).

Victorero, Annelt. 2015. Available at: www.wider.unu.edu/publication/joseph-stiglitz-and-amartya-sen-sustainable-development-goals (Accessed on 12 December 2018).

Walsen, Tariq, and Fleishman, Rachel. 'How War Games Can Help South Asia Respond to Climate Change.' *The Diplomat,* March 26, 2018. Available at: www.thediplomat.com/2018/03/how-war-games-can-help-south-asia-respond-to-climate-change (Accessed on 12 December 2018).

Williams, C. C., and Millington, A.C. 'The Diverse and Contested Meanings of Sustainable Development.' *The Geography Journal*, 170(2) (2004).

York, R., and Dunlap, R.E. 'Environmental Sociology.' In George Ritzer (Ed.), *Sociology*. Hoboken, New Jersey: Wiley Blackwell, 2016.

11 Technological developments for monitoring/forecasting extreme weather over South Asia

Sanjib Bandyopadhyay and Ganesh Kumar Das

The South Asian region is prone to extreme weather. Every year, hundreds of people die due to tropical cyclones (TCs), thunderstorms with lightning and heavy rainfall associated with floods, along with heat/cold waves. More than 200 people died due to a severe thunderstorm/*aandhi* (wind storm in the local Hindi language) over Delhi, Haryana and Uttar Pradesh in May 2018. Recently, some of the deadliest floods have been observed in the states of West Bengal, Jammu & Kashmir, Uttarakhand and Kerala and in the major cities of Mumbai, Gurugram, Bengaluru and Chennai. Coastal states of India are severely affected by very severe cyclonic storms (VSCSs), for example, Phailin in Odisha (2013), VSCS Hudhud (2014) in Andhra Pradesh, VSCS Vardah (2016) in Tamil Nadu, VSCS Ockhi (2017) in Kerala and most recently VSCS Titli (2018) in Odisha/Andhra Pradesh. The latest addition to this list is the super cyclonic storm Amphan that hit West Bengal on 20 May 2020, and the disaster caused by it has superseded almost all previous records in the 21st century. Heat waves during summer (April–June) kill many in Pakistan and North/Central India, whereas fog during the winter season (January–February) severely disrupts transport in general and aviation in particular. It has been observed that in this era of climate change, the intensity of extreme weather has increased. Stronger or more frequent weather extremes are likely to occur under climate change, such as more intense downpours and stronger cyclonic winds. Improved weather prediction, therefore, will be vital to give communities more time to prepare for dangerous storms and will save lives and minimize the damage to infrastructure.

The main objective of this chapter is to illustrate the major techniques and instruments of forecasting severe weather to reduce the loss of life and property and to promote sustainable development in the South Asian Region. This chapter discusses the technological development for monitoring/forecasting severe weather events in India. The study emphasizes the role of the Regional Specialized Meteorological Centre (RSMC), New Delhi, for monitoring and forecasting tropical cyclones over the North Indian Ocean

(NIO) and coordinating with the member states of South Asia, along with the utilization of satellite, radar and numerical weather prediction (NWP) in monitoring tropical cyclones and nor' westers (severe thunderstorms) over eastern India. Analysis of different forecasting techniques that save lives of hundreds of people during tropical disturbances in this region are also dealt with. Also, modes of dissemination of weather advisories and warnings are discussed in detail, with a special mention of Amphan and coping with contemporary climate change.

New communication networks and innovative systems and technology for forecasting (e.g. high-speed internet, wireless communication, digital climate databases, synergic workstations, nowcasting systems, ensemble forecasting systems) have emerged which provide the opportunity to improve weather prediction services, particularly in India. These innovations ensure the accurate provision of forecasts and warnings in graphical formats. The meteorological services provided by the Indian Meteorological Department to different sectors of the economy are presented in Figure 11.1.

As the figure shows, forecasting and weather prediction services are very useful to different departments and economic sectors apart from the general public. The tourism, aviation and power sectors, in addition to agriculture, shipping, fisheries and non-conventional energy sources,

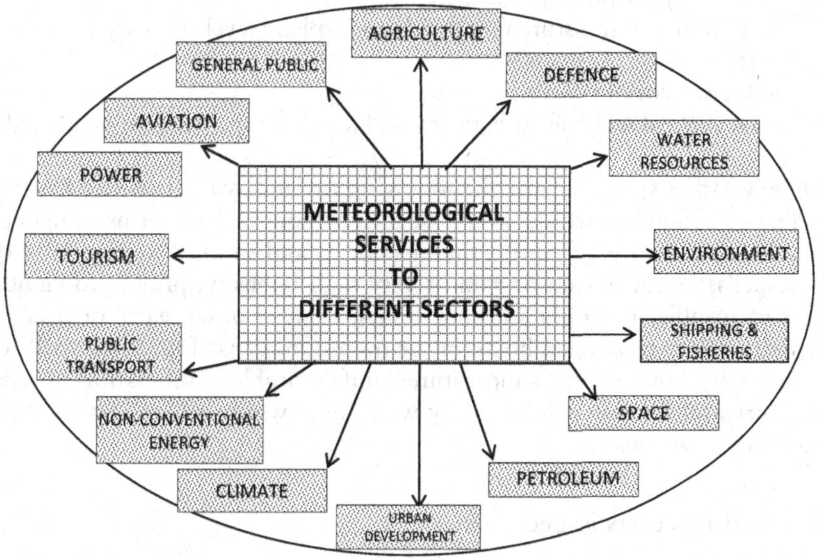

Figure-11.1 Activity of Indian Meteorological Department (IMD)

Source: http://www.imdkolkata.gov.in

coordinate their activities with these weather predictions. As a matter of fact, accurate prediction is extremely important for successful functioning of other activities.

It is worth mentioning here that the major weather systems affecting the region are mainly of four crucial categories: tropical cyclones, monsoon systems, local severe storms and western disturbances. Among these, the most aggressive weather system that causes extensive damage to lives and property are tropical cyclones. Others have usual courses of action. Tropical cyclones are warm-core low-pressure systems with a large vortex in the atmosphere, which is maintained by the release of latent heat by convective clouds over warm oceans. Typically, their diameters at the surface range between 100 and 1000 km, and the vertical extent is about 10 to 15 km, with the axis tilting towards colder regions. In the Northern hemisphere, the winds in a cyclone blow anticlockwise in the lower troposphere and clockwise in the upper troposphere. However, in the southern hemisphere, the winds of cyclones blow in the opposite direction. A fully developed cyclonic storm has four major components of horizontal structures: the eye, eyewall cloud region, rain/spiral bands and outer storm area. Tropical disturbances are convective in nature and migratory in character. The formation of tropical cyclones depends on the following conditions that are necessary but not sufficient:

- Warm sea surface temperature greater than 26°C.
- Potentially unstable atmosphere.
- Availability of moisture up to mid-troposphere levels (5.8 km above sea surface).
- Sufficient coriolis force.
- Low vertical wind shear between surface and upper troposphere winds.

These weather systems cause the following variant weather-related hazards in this area: floods, earthquakes, droughts, tornadoes, hailstorms, tsunamis, fog, heat/cold waves, frequent thunderstorms and so on. Prolonged dry or wet spells, untimely rain or rain of very high intensity, prolonged cloudiness or insufficient bright sunshine hours, delayed onset, early or delayed withdrawal or prolonged break in the monsoon, dense fog, frost, spells of abnormally low or high temperatures, unfavourable combination of temperature and humidity, flow of dry wind, high wind and so on also affect economic activities.

Types of forecasts issued

Various types of forecasts are generated by the Regional Specialized Meteorological Centre, Indian Meteorological Department, to cater to diverse purposes that can be summarized in the following manner.

Type of forecast	*Validity*
Nowcasting	Few hours
Short-range forecast	1–3 days
Medium-range forecast	3–10 days
Extended-range forecast	10–30 days
Long-range forecast	Month to season
Climate predictions	Annual-decadal-century

Nowcasting

Current weather and forecasts up to a few hours ahead are given by this type of forecasting method. A nowcast is given mainly for severe weather like squalls, hailstorms and heavy rainfall during the next 2 to 3 hours based on various Doppler weather radar (DWR) products, model outputs, satellite products, lighting data and swirl products. An important strength of a nowcasting system is its ability to rapidly generate warning products and disseminate them in a variety of formats.

Short-range forecast (1–3 days)

The weather in each successive 24-hour interval is predicted up to 1–3 days by short-range forecasts. This forecast range is mainly concerned with the weather systems observed in the latest weather charts and numerical weather prediction products.

Medium-range forecasts (4 to 10 days)

Average weather conditions and the weather on each day may be prescribed with progressively fewer details and accuracy than that for short-range forecasts. The department is more dependent on numerical weather prediction products for issuing medium-range forecasts.

Long-range/extended-range forecasts (more than 10 days to a season)

There is no rigid time span for long-range forecasting, which may range from 1 month to an entire season. IMD, for example, is presently giving long-range forecasts of monsoon season in India (June to September).

Extended-range forecast products

Tools for forecasting extreme weather have advanced in recent decades. The following weather forecasting tools are of frequent and essential use.

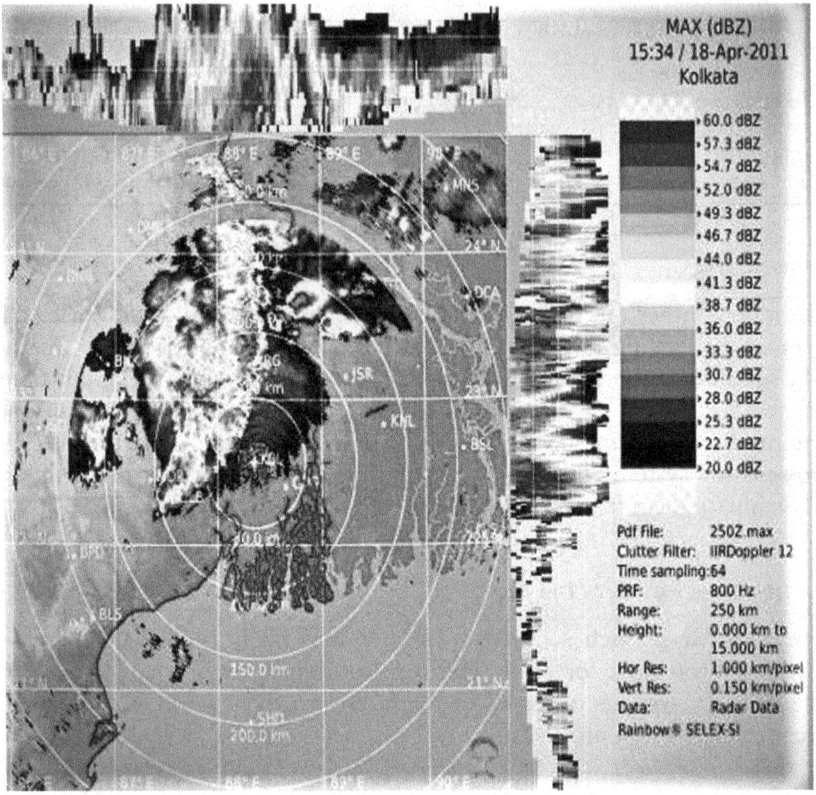

Figure 11.2 Radar (a) and satellite (b) pictures (IMD) for nowcasting

Source: http://www.imdkolkata.gov.in/

Surface observation/automated surface-observing systems

Surface observation/automatic weather stations (AWSs) and automatic rain gauge stations (ARGs) constantly monitor weather conditions on the Earth's surface. They report hourly data about precipitation, temperature and wind and so on. The observational data are essential for improving forecasts and warnings.

GPS sonde observation

Global positioning system (GPS) sondes are our primary source of upper-air data. At least twice per day, radio sondes are launched throughout the country and at all the airports. In its two-hour trip, the radio sonde floats up to the upper stratosphere, where it collects and sends back data every second

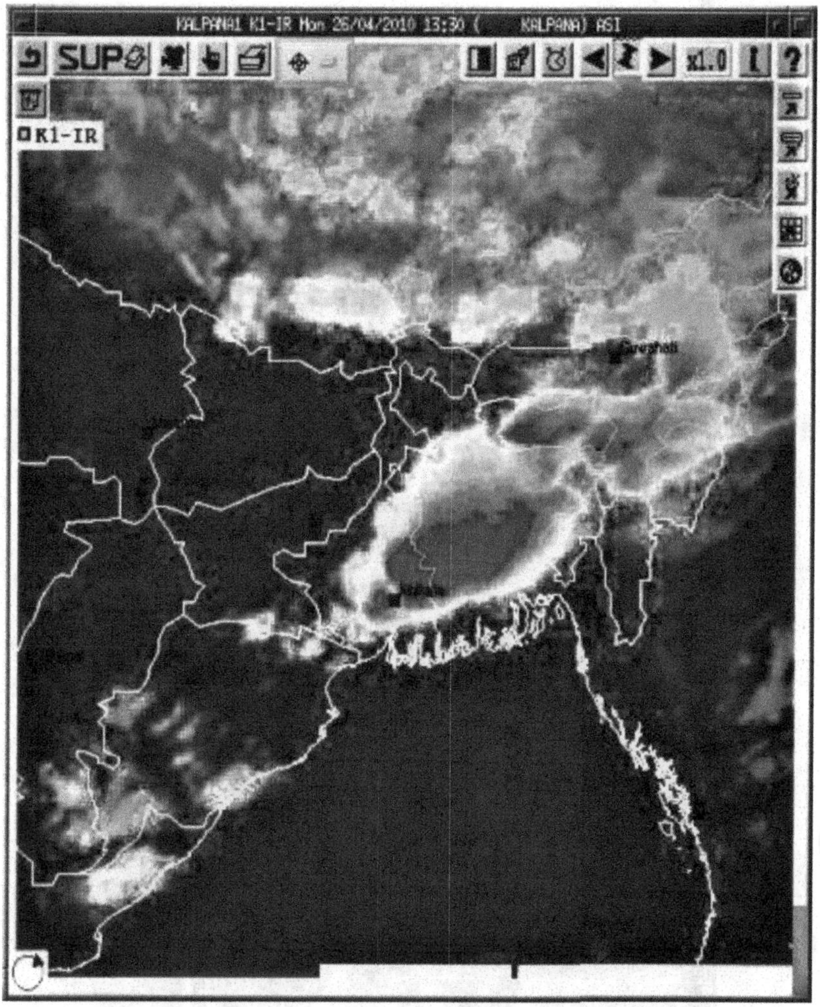

Figure 11.2 Continued

about air pressure, temperature, relative humidity, wind speed and wind direction. During severe weather, frequent observations are taken.

Synergie workstation

This is a state-of-the-art workstation for analysing and forecasting weather. At this workstation, one can see the entire real-time weather chart like synoptic charts, upper air charts, auxiliary charts, satellite pictures and various numerical weather products for weather forecast and warning. This

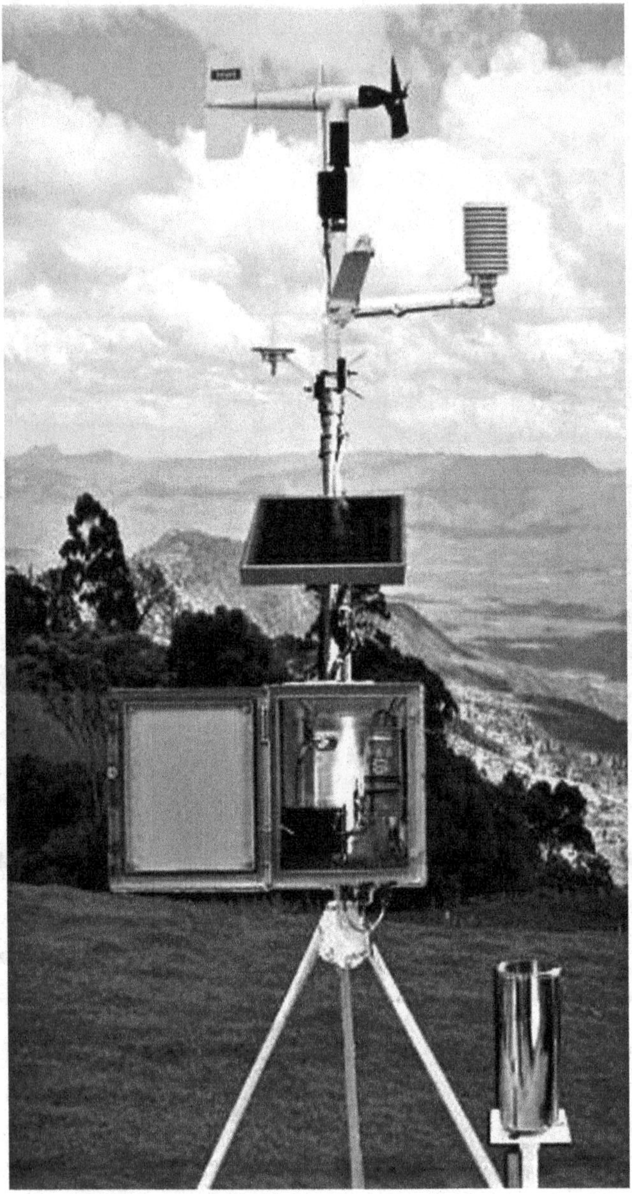

Figure 11.3 Automatic weather station

Source: http://www.imdkolkata.gov.in/

Figure 11.4 GPS balloon

Source: http://www.imdkolkata.gov.in/

platform has the facility to analyse different weather charts and superimpose one parameters/satellite picture over other parameters. The final forecast can be disseminated per requirements. The products used by the Synergie workstation are described in the following.

Synergie products

1 Satellite

Satellite is one of the most important tools to monitor the development; intensification and movement of a tropical cyclone when it is in the deep sea. It covers the whole data-sparse region over the ocean. There are various derived satellite products available for analysis and forecasting of all scale forecasts. Satellites provided by all the global Meteorological Organization/Institute are being used to monitor severe weather over South Asia. Some of the important satellites are Himawari from Japan Meteorological Agency (JMA) and Eumetsat. Imd (from India) that monitor severe weather through INSAT 3D and INSAT 3DR. Various types of channel/satellite products are used to predict the development and intensity of tropical cyclones, thunder squalls and heavy rainfall events. Moreover, the centre/eye of cyclone can be gauged by satellite picture. Assorted derived products like cloud top temperatures, cloud motion vector (CMV) wind, outgoing long-wave radiation (OLR), divergence-convergence patterns of lower and upper levels, wind shear tendency and so on are used for day-to-day forecasting. Products from both geostationary and polar orbiting satellites are used for different wavelengths.

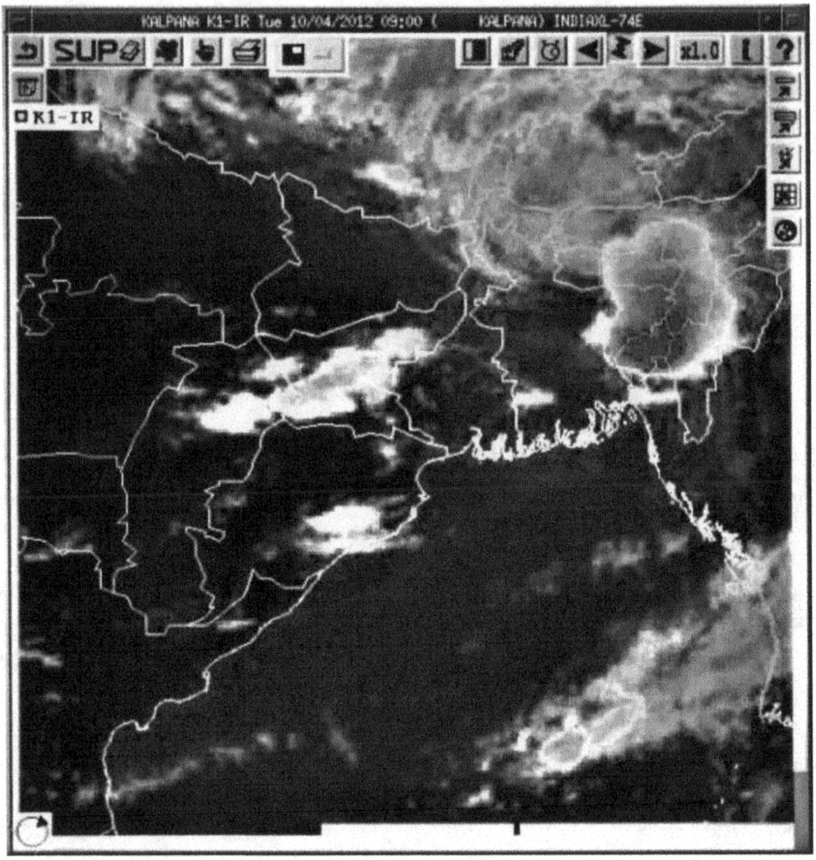

Figure 11.5 Satellite product

Source: http://www.imdkolkata.gov.in/, http://foreignsat.imd.gov.in

2 *Doppler weather radar*

DWR is the best tool for nowcasting of thunderstorms, heavy rainfall and flash floods when severe weather like thunder squalls and tropical cyclones come under the range of DWR. Various DWR products like Max (Z), PPI and VVP are used for heavy rainfall/thunderstorm forecasting. Dual polarization technology allows forecasters to differentiate between types of precipitation and amounts. DWR helps to produce more accurate forecasts.

3 *Numerical weather prediction*

The technology of weather forecasting has become mostly dependent on software skills to solve the equations of the structural models that govern

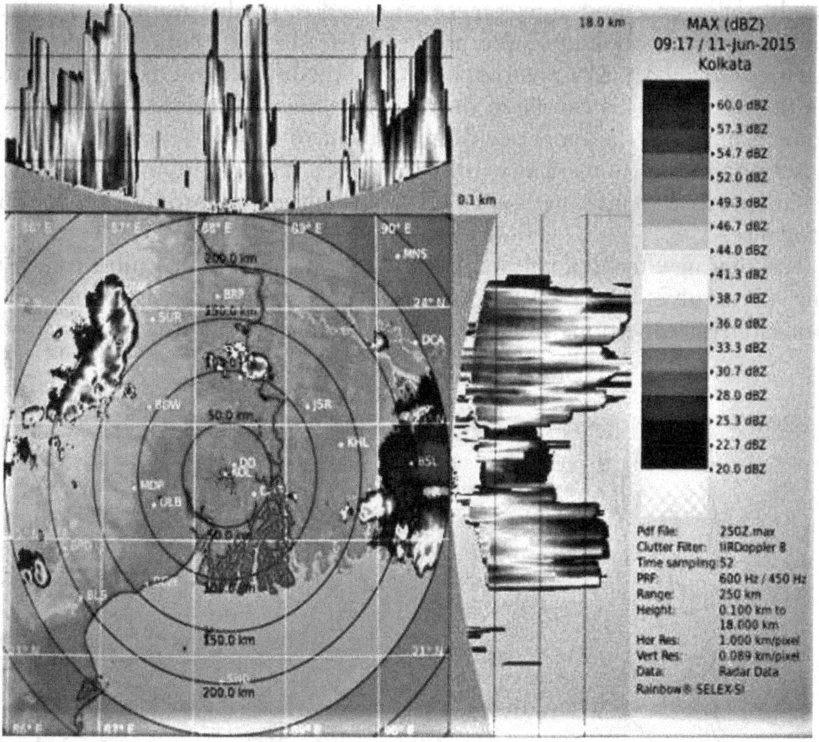

Figure 11.6 Max (Z) from DWR Kolkata

Source: http://www.imdkolkata.gov.in/

atmospheric processes – this is known as numerical weather prediction. The basic idea of numerical weather prediction is to sample the state of the fluid at a given time and use the equations of fluid dynamics and thermodynamics to estimate the state of the fluid at some time in the future. Models are initialized using the observed data. Irregularly spaced observations are processed by data assimilation and objective analysis methods, which perform quality control and obtain values at locations usable by the model's mathematical algorithms. The data are then used in the model as the starting point for a forecast. The set of equations used to predict what are known as the physics and dynamics of the atmosphere are called primitive equations. These equations are initialized from the analysis data, and their rates of change are determined. The rates of change predict the state of the atmosphere for a short time into the future. The equations are then applied to this new atmospheric state to find new rates of change, and these new rates of change predict the atmosphere at a yet further time into the future. This time-stepping procedure is continually repeated until the solution reaches the desired forecast time. Various NWP products, like vortices at different levels, divergence and convergence, are used to predict rainfall on a day-to-day basis. The visual output produced by a model solution is known as a *prognostic chart*. The raw output is often modified before being presented as the forecast. This can be in the form of statistical techniques to remove known biases in the model or of adjustment to take into account consensus among other numerical weather forecasts. Model output statistics is a technique used to interpret numerical model output and produce site-specific guidance.

Another new technology that has become available is ensemble forecasting – it involves using NWP not to produce a single forecast of the future atmospheric conditions but rather many different solutions. The various

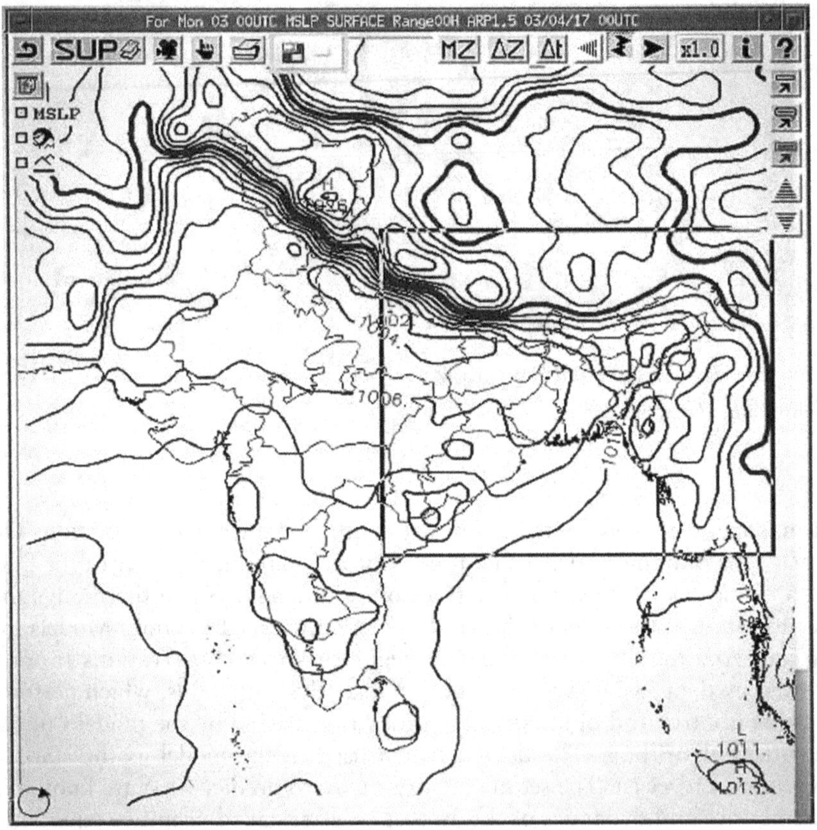

Figure 11.7 Various NWP products
Figure 11.7 (a) Mean sea level pressure (MSLP)
Source: http://www.imdkolkata.gov.in/

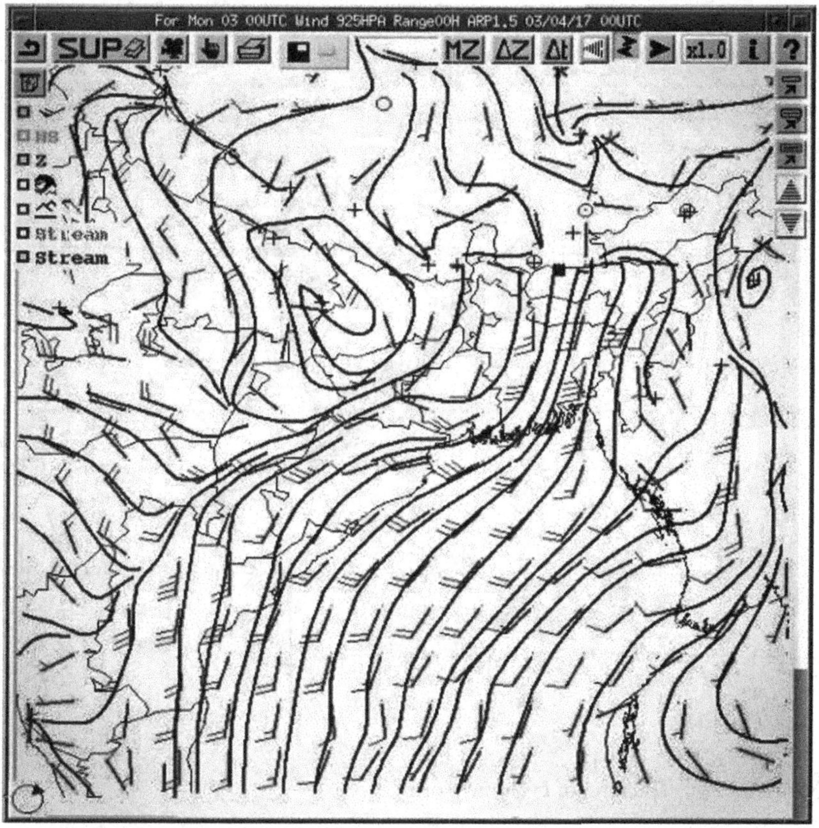

Figure 11.7(b) Low-level wind and streamline

different forecasts are derived by using slightly changed initial data. This provides an "ensemble" of possible future conditions and offers some additional information about how likely the weather may be to change in a particular way. This ensemble method gives us a way to estimate the uncertainty in our forecasts quantitatively.

4 Geographic information systems and global positioning system

Geographic information systems (GIS) are designed for capturing, storing, analysing and managing data and associated attributes which are spatially referenced to the Earth. The global positioning system provides location-specific information. Together, GIS and GPS provide a powerful

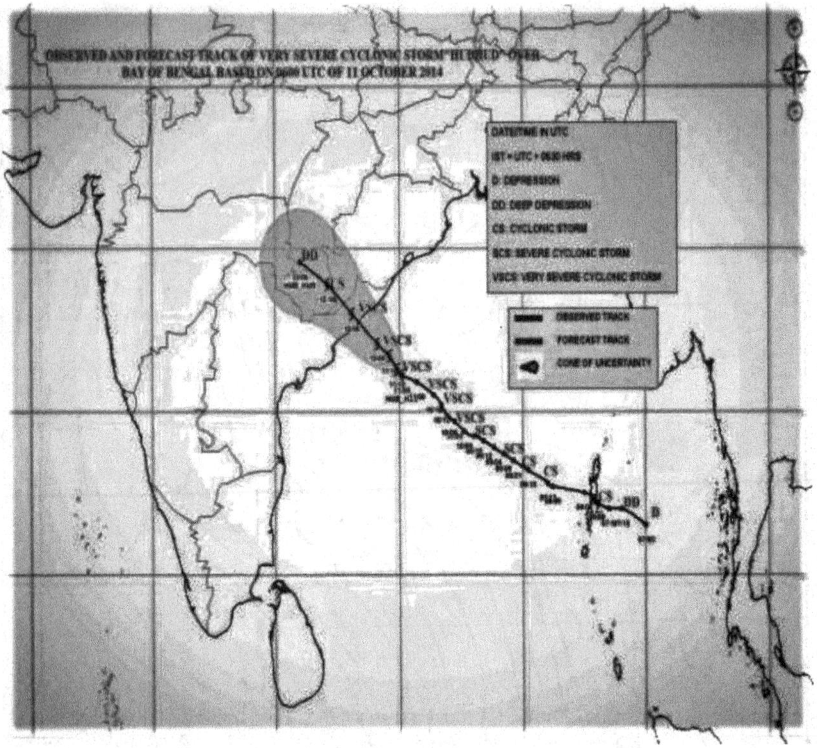

Figure 11.7(c) Tracking of cyclonic storm

technological tool to effectively deliver user- and location-specific warnings and forecasts.

5 Supercomputers

Supercomputers collect, process and analyse billions of observations from weather satellites, weather balloons, buoys and surface stations from around the world. The supercomputer system is the backbone of modern forecasting. With huge computing capacity, it can process quadrillions of calculations per second. Supercomputers are almost 6 million times more powerful than the average desktop computer. Observational data collected by Doppler radar, radio sondes, weather satellites, buoys and other instruments are fed into computerized numerical forecast models. The models use equations, along with new and past weather data, to provide forecast guidance to meteorologists.

Forecasting methodology

Weather forecasting is the application of science and technology to predict the conditions of the atmosphere for a given location and time. Weather forecasts are made by collecting quantitative data about the current state of the atmosphere at a given place and using meteorology to project how the atmosphere will change. Weather forecasting now relies on computer-based models that take many atmospheric factors into account. Human input is still required to pick the best possible forecast model, which involves pattern recognition skills, teleconnections, knowledge of model performance and knowledge of model biases. Inaccuracy of forecasting occurs due to the chaotic nature of the atmosphere and often the inadequacy of massive computational power required to solve the equations that describe the atmosphere, the error involved in measuring the initial conditions and an incomplete understanding of atmospheric processes. The effective use of ensembles and model consensus helps to minimise error and pick the most likely outcome.

So the three important forecasting steps include first assessing the present state, called the 'analysis'; then predicting a future state by running a technically designed model; and, finally, interpreting the model results, called a 'prognostic chart', given forecasting experience.

Major steps in the forecast process

- Data collection
- Quality control
- Data assimilation
- Model integration
- Post-processing of model forecasts
- Human interpretation
- Product and graphics generation

Weather is observed throughout the world and the data are distributed in real time. Weather-related data are collected from different sources like surface observations, GPS/radio sondes and radar profilers, fixed and drifting buoys, ship/aircraft observations, satellite soundings/imagery, cloud and water vapour track winds and so on. After collecting data from variant sources and controlling their quality, all the data are assimilated to incorporate into the best-fit model. The model forecasts using the techniques already mentioned in the last section after processing the data. Finally, the meteorologists interpret the results of the model outcomes and disclose in the proper channels with the support of appropriate products and graphics generation.

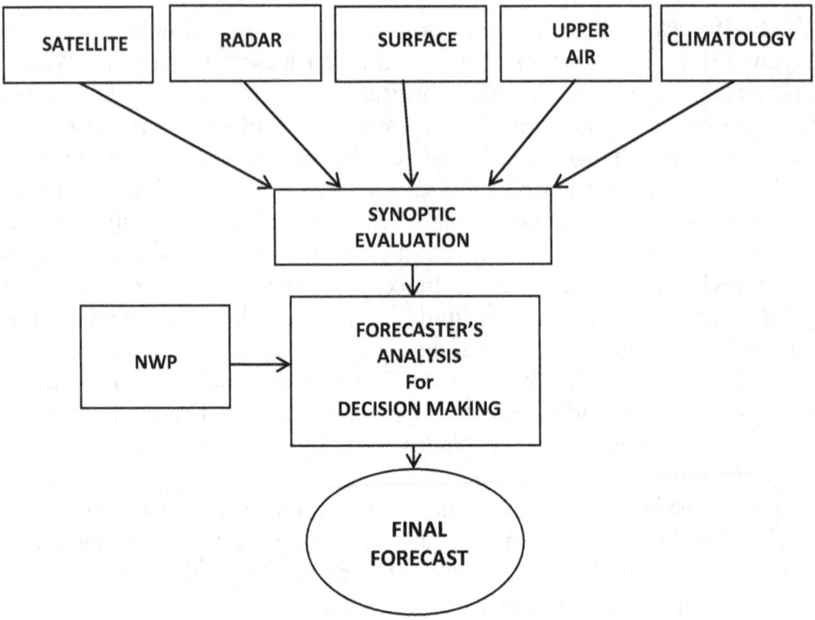

Figure 11.8 Various tools/products for ensemble forecasting

Source: http://www.imdkolkata.gov.in/

Some important kinds of forecasting

Monitoring/nowcasting of severe thunderstorms

The forecasting of mesoscale phenomena has been a challenge to atmospheric scientists, since imposing boundary conditions on mathematical equations of those phenomena (e.g. thunderstorms, squalls) is quite difficult due to the comparable values of zonal and meridional components. Broadly, four ingredients have been found essential for thunderstorm activity: instability, wind shear, lifting and moisture. The following tools are used to monitor severe thunderstorms in India.

1 Synoptic chart/upper air chart
2 Thermodynamic parameters
3 Satellite picture
4 Radar
5 NWP
6 Short-range warning of intense rainstorms in localized systems (SWIRLS)

A real-time SWIRLS system uses a number of algorithms to derive storm motion vectors. These include tracking of radar echoes (TREC) and multi-scale optical flow by variational analysis (MOVA). The MOVA algorithm uses optical flow, a technique commonly used in motion detection in image processing, and variational analysis is used to derive the motion vector field. By cascading through a range of scales, MOVA can better depict the actual storm motion vector field as compared with TREC, which does well in tracking small-scale features and storm entities. Nowcast system SWIRLS provide forecasts for intense rainfall events for the next 2 hours based on the latest radar data.

Monitoring/forecasting of tropical disturbances/cyclones

Tropical cyclonic storms are large revolving vortices with a ring of strong winds around a warm core extending horizontally from 150–1000 km and vertically to about 12–14 km above sea level, causing enormous damage to life and property at the time of crossing the coast and subsequent movement over the land. Cyclonic disturbances formed over the North Indian Ocean are categorized according to their maximum sustained wind speeds. The Regional Specialized Meteorological Centre has been assigned the responsibility of issuing Tropical Weather Outlooks and Tropical Cyclone Advisories for the benefit of the countries in the World Meteorological Organisation (WMO)/United Nations Economic and Social Commission for Asia-Pacific (ESCAP) Panel region bordering the Bay of Bengal and the Arabian Sea, namely Bangladesh, Maldives, Myanmar, Oman, Pakistan, Sri Lanka and Thailand.

Tropical weather outlooks

The RSMC has been assigned the responsibility of issuing Tropical Weather Outlooks for the North Indian Ocean for cyclogenesis for 2 successive weeks in three different categories with varying probability of cyclogenesis (formation of depression or higher intensity).

- Low (1–33% probability)
- Moderate (34–67% probability)
- High (68–100% probability)

The outlook issued is based on the position and amplitude of the Madden Julian oscillation (MJO) index and its forecast position in the subsequent 2 weeks. The outlook confirms whether the situation is favourable for enhancement of convective activity over the North Indian Ocean region. The extended-range forecast by the Ministry of Earth Sciences (MoES) Climate Forecast System (CFS) model is also taken into consideration while doing this. The genesis potential parameter based on the IMD global forecast system (GFS) is also considered for development of potential zone for cyclogenesis for the next 7 days, along with other numerical models like Indian Meteorological Department's Global Forecast System (IMDGFS), National Centers for Environmental Prediction's Global Forecast System (NCEP GFS), Global Ensemble Forecasting System (GEFS), National Center for Medium Range Weather Forecast (NCMRWF's) Unified Modeling (NCUM) system, NCMRWF global Ensemble Prediction System (NEPS), The European Centre for Medium-Range Weather Forecasts (ECMWF) and so on.

As India is surrounded by the Arabian Sea, Bay of Bengal and Indian Ocean, the frequency of cyclonic storms is naturally high here, and the phenomenon of global warming and consequent rise in sea level makes the country more prone to such storms. As a matter of fact, it is imperative to discuss here in more detail the most crucial kind of warning that saves numerous lives from being shattered by nature's wrath, that is, cyclone warnings.

Cyclone forecasts/warnings

Cyclone monitoring by satellites is done with the help of very high-resolution radiometers, working in the visual and infrared regions of the spectrum to obtain an image of the cloud cover and its structure. When the system is within the radar range, it is monitored by weather radars installed along the

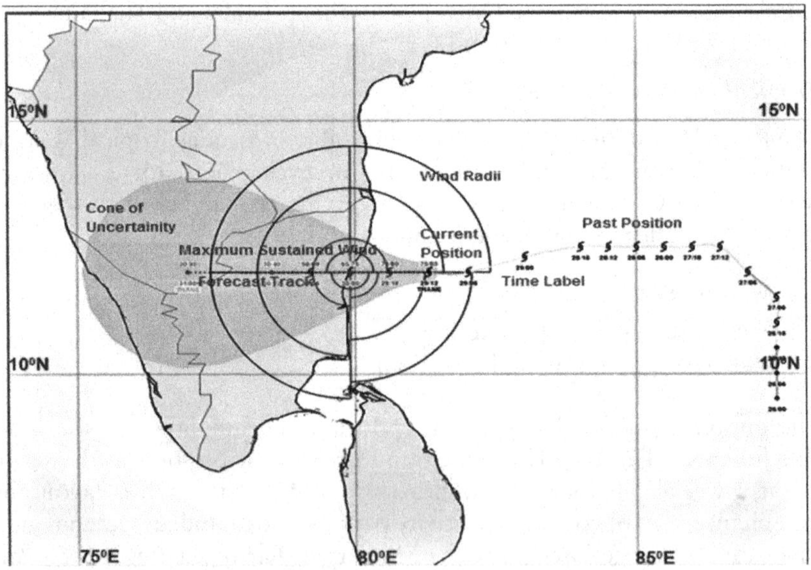

Figure 11.9 Track forecasting and cone of uncertainty of tropical cyclone

Source: http://www.imdkolkata.gov.in/

coastal region. There are networks of weather stations monitoring pressure fall and wind velocities along the coastal stations. High-resolution models are run to forecast the track and intensity of the cyclone, including probable landfall point and time of landfall, estimated wind speed and storm surge if any.

Four-stage warning

Cyclone forecasting or warning is, generally, executed in four stages, as described subsequently.

i Pre-Cyclone watch

A Pre-Cyclone Watch is issued to senior disaster management officials indicating the formation of a tropical disturbance that has a potential tendency to intensify into a cyclone and the part of the coastal area that is likely to be affected.

ii Cyclone alert

This is issued at least 48 hours in advance, indicating expected adverse weather conditions.

iii Cyclone warning

This is issued at least 24 hours in advance, indicating the latest position of the tropical cyclone, intensity, time and point of landfall, storm surge height, types of damage expected and action suggested.

iv Post-landfall outlook

Issued about 12 hours before landfall and until cyclone force winds prevail. District collectors of interior and remote areas are duly informed.

De-warning

De-warning messages are issued when the tropical cyclone weakens into the depression stage.

A recent case study on cyclone forecasting in the Bay of Bengal

Supercyclone Amphan

The most powerful and deadly cyclonic storm, which eventually culminated in a super cyclonic storm (SuCS) and hit the two eastern states of India and part of Bangladesh on 20th May, 2020, was named Amphan. Amphan (pronounced as UM-PUN) was the first SuCS over the Bay of Bengal after the Odisha SuCS of 1999. This SuCS was very accurately predicted and forecast by the Indian Meteorological Department, which enabled the administration to minimize the loss of lives, if not property. The successive stages of formation of Amphan as monitored by IMD and its precise forecasting techniques are presented in the following sections.

Successive stages of formation of Amphan

This section portrays the successive stages of formation, movement and landfall of Amphan over Bay of Bengal and West Bengal (eastern Indian state) land territory that will help to evaluate its severity.

- It originated from the remnant of a low-pressure area which occurred in the near Equatorial easterly wave over the south Andaman Sea and adjoining the southeast Bay of Bengal (BoB) on 13 May.
- It concentrated into a depression (D) over the southeast BoB in the early morning of 16 May and further intensified into a deep depression (DD) the same afternoon.
- It moved north-north westwards and intensified into a cyclonic storm over the southeast BoB in the evening of 16 May 2020.
- Moving nearly northwards, it further intensified into a severe cyclonic storm (SCS) over the southeast BoB the morning of 17 May.
- It underwent rapid intensification during the subsequent 24 hours and accordingly intensified into a very severe cyclonic storm by the afternoon of the 17th, an extremely severe cyclonic storm (ESCS) in the early hours of the 18th and a SuCS around noon of 18 May 2020.
- It maintained the intensity of SuCS over the west-central BoB for nearly 24 hours before weakening into an ESCS over the west-central BoB around noon of 19 May.
- Thereafter, it weakened slightly and crossed the West Bengal–Bangladesh coasts as a VSCS, across Sundarbans, during 1530–1730 hrs IST of 20 May, with maximum sustained wind speed of 155–165 kmph, gusting to 185 kmph.
- It lay over West Bengal as a VSCS, gradually moving north-north eastwards during the late evening to night of 20 May.
- It moved very close to Kolkata, the eastern metropolis of India, during this period.
- Moving further north-north eastwards, it weakened into an SCS over Bangladesh and adjoining West Bengal around midnight of 20 May.

It weakened further into a CS over Bangladesh in the early hours of 21 May, into a DD over Bangladesh around noon of 21 May and into a D over north Bangladesh in the evening of the same day.

It further weakened and lay as a well-marked low pressure area over north Bangladesh and the neighbourhood around midnight of 21 May.

Monitoring of the Amphan cyclone

The super cyclone Amphan was monitored with the help of available satellite observations:

- INSAT 3D and 3DR.
- Polar orbiting satellites, including SCATSAT, ASCAT and so on.
- Available ship and buoy observations in the region.
- From 18 May midnight (1800 UTC) onwards to 20 May, the system was tracked gradually by IMD Doppler weather radar at Visakhapatnam, Gopalpur, Paradip, Kolkata and Agartala as it moved from south to north.
- IMD also utilized DWR products from the DRDO Integrated Test Range, Chandipur, Balasore, for tracking the system.
- Various numerical weather prediction models run by Ministry of Earth Sciences (MoES) institutions (IMD, IITM, NCMRWF & INCOIS)
- Various global models and IMD's dynamical-statistical models developed inhouse were utilized to predict the genesis, track, landfall and intensity of the cyclone.
- A digitized forecasting system of IMD was utilized for analysis and comparison of guidance from various models, decision-making processes and warning product generation.

Forecast performance of Amphan

i Genesis forecast

- The system was monitored from 23 April about 3 weeks prior to the formation of a low pressure area over the southeast BoB on 13 May.
- In the extended range outlook issued on 7 May, cyclogenesis (formation of depression) was predicted with low probability in the later part of the week during 8–14 May 2020. It was also predicted that the system would intensify further and move initially north-north westwards and recurve north-north eastwards thereafter towards the north BoB.
- In the Tropical Weather Outlook issued on 9 May, it was indicated that a low-pressure area would form over the region on 13 May (96 hours prior to formation of the system) under the influence of the remnant cyclonic circulation persisting over the region during 6–12 May.
- In the Tropical Weather Outlook issued on 11 May, it was indicated that cyclogenesis (formation of depression) would occur around 16 May (48 hours prior to formation of the low-pressure area and 120 hours prior to formation of depression) over the BoB. The low-pressure area

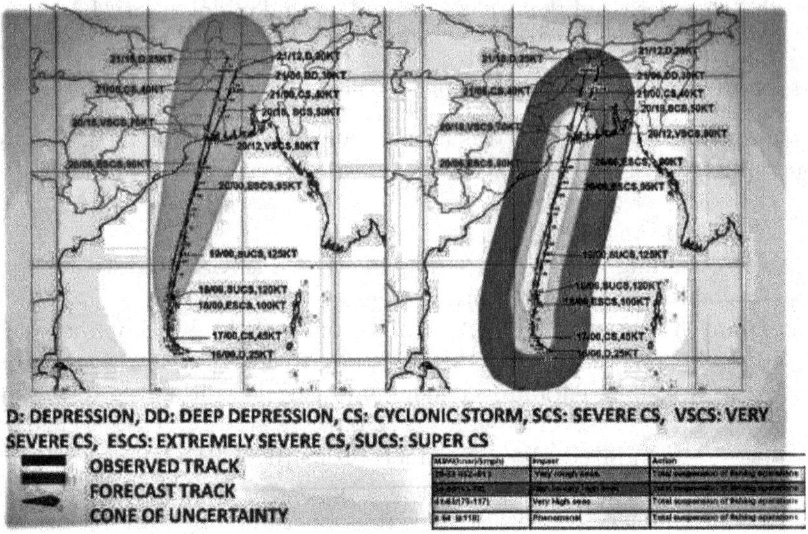

Figure 11.10 Forecast track and cone of uncertainty for Amphan

Source: imd.gov.in

formed on 13 May and concentrated into a depression on the morning of 16 May.

ii Track, intensity and landfall forecast

- First information was provided in the extended range outlook issued on 7 May (about 6 days prior to formation of LPA, 9 days prior to formation of depression and 13 days prior to landfall) and indicated that the system would intensify into a cyclonic storm and move initially north westwards and re-curve north-north eastwards towards the north BoB.
- In the Tropical Weather Outlook, press release and informatory message to the government of India issued on 13 April (on the day of development of LPA, 3 days prior to formation of the depression and 7 days prior to landfall), it was indicated that the system would intensify into a cyclonic storm by the evening of the 16th and would move initially north westwards until the 17th and then re-curve north-north eastwards towards the north BoB.
- Actually, the depression formed in the morning (0000 UTC) of the 16th, the cyclonic storm in the evening (1200 UTC) of the 16th and the system moved north-north westwards until the evening of the 17th (1200 UTC), followed by north-north eastward re-curvature thereafter and crossed the West Bengal coast in the afternoon of the 20th.

- In the first bulletin issued at 0845 IST of 16th May (104 hrs prior to landfall) with the formation of the depression, it was indicated that the system would intensify into a cyclonic storm and move north-north westwards until 17 May, followed by north-north eastward re-curvature towards the West Bengal coast during 18–20 May and cross the West Bengal coast with maximum sustained wind speeds of 155–165 kmph, gusting to 180 kmph.
- In the bulletin issued at 1645 hrs IST on 16 May (24 hrs prior to rapid intensification), rapid intensification of the system was predicted, and the system rapidly intensified from the afternoon of the 17th onwards.
- In the bulletin issued at 0845 hrs IST on 17 May (80 hrs prior to landfall), it was precisely mentioned that the system would cross the West Bengal-Bangladesh coasts between Sagar Island (West Bengal) and Hatiya Islands (Bangladesh coast) during afternoon to evening of 20 May with maximum sustained windspeeds of 155–165 kmph, gusting to 185 kmph. The predicted track indicated landfall across Sunderbans on the afternoon of the 20th.
- IMD continuously predicted from 16 May that Amphan would cross the West Bengal coast as a very severe cyclonic storm with wind speeds of 155–165 kmph, gusting to 180 kmph on 20 May.

Cyclone warnings for Amphan

- In the first bulletin released at 0845 hrs IST on 16 May (104 hrs prior to landfall), a Pre-Cyclone Watch for the West Bengal-north Odisha coasts was issued.
- The warnings were further upgraded and a Cyclone Watch for the West Bengal and north Odisha coasts was issued at 2030 hrs IST on 16 May (92 hrs prior to landfall).
- Cyclone Alert (Yellow Message) for the West Bengal and north Odisha coasts was issued at 0840 hrs IST on 17 May (80 hrs prior to landfall).
- Cyclone Warning (Orange Message) for the West Bengal and north Odisha coasts was issued at 0845 hrs IST on 18 May (56 hrs prior to landfall).
- Post-landfall outlook (Red Message) for the interior districts of Gangetic West Bengal, Assam and Meghalaya was issued at 2330 hrs IST on 19 May (17 hrs prior to landfall).

Thus, IMD accurately predicted the landfall point and time, track and intensity as well as the occurrence of adverse weather like wind, rainfall and storm surge associated with the fury of the cyclone Amphan.

With these instances of technological progress and improvements in communication, revolution in computing ability and numerical weather predictions, as well as with a better understanding of physics of the atmosphere, the Weather Forecast Office will be able to predict severe weather and associated hazards more accurately with less probability of uncertainties and with longer lead times. Severe weather outlooks will extend beyond five days, tropical cyclone outlook and forecasts beyond 7 days, and the threat of floods will be known weeks in advance resulting in less loss of life and property. In a nutshell, given the considerable technological development and advances in the monitoring and forecasting of extreme weather conditions in South Asia, millions of people can be moved to safer places before cyclones hit the coasts, storms start swirling or clouds burst, if not before an earthquake, thus minimizing the loss of life. It is important for us to trust the power of scientists, technologists and forecasters to restrain the harms of climatic disasters to save humanity and its habitat.

Index

Printed in the United States
by Baker & Taylor Publisher Services